Evolutionary Economics and Social Complexity Science

Volume 17

The Japanese Association for Evolutionary Economics (JAFEE) always has adhered to its original aim of taking an explicit "integrated" approach. This path has been followed steadfastly since the Association's establishment in 1997 and, as well, since the inauguration of our international journal in 2004. We have deployed an agenda encompassing a contemporary array of subjects including but not limited to: foundations of institutional and evolutionary economics, criticism of mainstream views in the social sciences, knowledge and learning in socio-economic life, development and innovation of technologies, transformation of industrial organizations and economic systems, experimental studies in economics, agent-based modeling of socio-economic systems, evolution of the governance structure of firms and other organizations, comparison of dynamically changing institutions of the world, and policy proposals in the transformational process of economic life. In short, our starting point is an "integrative science" of evolutionary and institutional views. Furthermore, we always endeavor to stay abreast of newly established methods such as agent-based modeling, socio/econo-physics, and network analysis as part of our integrative links.

More fundamentally, "evolution" in social science is interpreted as an essential key word, i.e., an integrative and /or communicative link to understand and re-domain various preceding dichotomies in the sciences: ontological or epistemological, subjective or objective, homogeneous or heterogeneous, natural or artificial, selfish or altruistic, individualistic or collective, rational or irrational, axiomatic or psychological-based, causal nexus or cyclic networked, optimal or adaptive, micro- or macroscopic, deterministic or stochastic, historical or theoretical, mathematical or computational, experimental or empirical, agent-based or socio/econo-physical, institutional or evolutionary, regional or global, and so on. The conventional meanings adhering to various traditional dichotomies may be more or less obsolete, to be replaced with more current ones vis-à-vis contemporary academic trends. Thus we are strongly encouraged to integrate some of the conventional dichotomies.

These attempts are not limited to the field of economic sciences, including management sciences, but also include social science in general. In that way, understanding the social profiles of complex science may then be within our reach. In the meantime, contemporary society appears to be evolving into a newly emerging phase, chiefly characterized by an information and communication technology (ICT) mode of production and a service network system replacing the earlier established factory system with a new one that is suited to actual observations. In the face of these changes we are urgently compelled to explore a set of new properties for a new socio/economic system by implementing new ideas. We thus are keen to look for "integrated principles" common to the above-mentioned dichotomies throughout our serial compilation of publications. We are also encouraged to create a new, broader spectrum for establishing a specific method positively integrated in our own original way.

More information about this series at http://www.springer.com/series/11930

Jun Tanimoto

Evolutionary Games with Sociophysics

Analysis of Traffic Flow and Epidemics

Springer

Jun Tanimoto
Graduate School of Engineering Sciences
Kyushu University
Fukuoka, Japan

ISSN 2198-4204 ISSN 2198-4212 (electronic)
Evolutionary Economics and Social Complexity Science
ISBN 978-981-13-2768-1 ISBN 978-981-13-2769-8 (eBook)
https://doi.org/10.1007/978-981-13-2769-8

Library of Congress Control Number: 2018959370

This Springer imprint is published by the registered company Springer Nature Singapore Pte Ltd.
The registered company address is: 152 Beach Road, #21-01/04 Gateway East, Singapore 189721, Singapore

Preface

It is my great pleasure that the positive feedback received for my previous book, *Fundamentals of Evolutionary Game Theory and Its Applications*, has given me the chance to publish another book.

As I mentioned in the preface of the first book, I started studying evolutionary game theory and other applied mathematics under the umbrella of Operational Research (OR). I was experiencing some frustration because I realized that it would be difficult to pose any meaningful social prescriptions in order to solve the so-called environmental problems, even though I specialize in the field of building physics, urban climatology, and other environmental engineering and sciences. Furthermore, a human decision is often more important than technical problem-solving for significantly impacting various environmental problems. Hence, I decided to shift my expertise from heat transfer theory and fluid dynamics to *social physics* and developed the concept of a *human-environmental-social system.*

For several decades, social physics has been a subject of fascination for mathematicians, statistical physicists, theoretical biologists, and information scientists. This interest is due to the possibility that social physics may elucidate some aspects of puzzling social phenomena.

Some important scientific journals have also started featuring new social physics studies, for example, how animal species have obtained an attitude of social cooperation that do not benefit the individual and what triggers human beings to acquire "language" as a mutual communicating protocol.

This book first gives readers some fundamental knowledge of evolutionary game theory, which is requisite for their own scientific challenges.

Subsequently, two interesting and important applications derived from evolutionary game theory are provided: traffic flow analysis and vaccination game. Both applications have attracted much attention, since relieving urban traffic jam (which contributes to poor urban air quality and the wasting of energy) and preventing the pandemic of worldwide infectious diseases are widely regarded as two of the biggest issues to be addressed in modern society.

I would be most honored if this book somehow serves as a guidebook for readers seeking new transdisciplinary areas.

Fukuoka, Japan Jun Tanimoto

Acknowledgments

This book owes its greatest debt to my coworkers, who were my excellent students. Chapter 2 relies on the contributions of Mr. Satoshi Kokubo (Mitsubishi Electric Corporation), Mr. Ryo Matsuzawa (Daikin Industries, Ltd.), and Mr. Yoshiro Iwamura (IDOM Inc.), who also participated in other projects described in Chaps. 3 and 4. Dr. Eriko Fukuda (Sanyo-Onoda City University) put great effort into the field survey described in Sect. 3.1. Chapter 3 contains a series of results based on numerical experiments obtained by Mr. Kousuke Nakamura (West Nippon Expressway Company Limited). The theoretical framework of Chap. 4 was developed by Mr. Kazuki Kuga (DC Fellow of Japan Society for the Promotion of Science). I greatly appreciate all of their contributions.

Last but not the least, I am grateful to Dr. Prof. Yuji Aruka at Chuo University for providing me with the opportunity to publish this book.

Contents

About the Author

Jun Tanimoto The author was born in 1965 in Fukuoka, but he grew up in Yokohama. He graduated in 1988 from the Department of Architecture, Undergraduate School of Science and Engineering, Waseda University. In 1990, he completed his master's program, and in 1993, he earned his doctoral degree from Waseda University. He started his professional career as a Research Associate at Tokyo Metropolitan University in 1990, moved to Kyushu University and was promoted to Assistant Professor (Senior Lecturer) in 1995, and became Associate Professor in 1998. Since 2003 he has served as full-titled Professor and Head of the Laboratory of Urban Architectural Environmental Engineering. He served as a Visiting Professor at the National Renewable Energy Laboratory (NREL), USA; at the University of New South Wales, Australia; at Eindhoven University of Technology, Netherlands; and at Max Planck Institute for Evolutionary Biology, Germany. Professor Tanimoto has published numerous scientific papers in building physics, urban climatology, and statistical physics and is the author of books such as *Fundamentals of Evolutionary Game Theory and Its Applications* (Springer; ISBN: 978-4-431-54961-1) and *Mathematical Analysis of Environmental System* (Springer; ISBN: 978-4-431-54621-4). He was a recipient of the Award of the Society of Heating, Air-Conditioning, and Sanitary Engineers of Japan (SHASE), the Fosterage Award from the Architectural Institute of Japan (AIJ), the Award of AIJ, and the IEEE CEC2009 Best Paper Award. He is involved in numerous activities worldwide, including editor at several international journals including *Applied Mathematics and Computation*, *PLOS One* and *Journal of Building Performance Simulation*, among others; committee member for many conferences; and expert at the IEA Solar Heating and Cooling Programme Task 23. He is also an active painter and novelist, and has been awarded numerous prizes in fine art and literature. He has created many works of art and published several books. He specializes in scenic drawing with watercolors and romantic fiction. For more information, please visit http://ktlabo.cm.kyushu-u.ac.jp/.

Chapter 1
Sociophysics Approach to Modeling and Analyzing the Human–Environmental–Social System

This chapter presents a general introduction to the book. We discuss why the concept of sociophysics, one of the upcoming interdisciplinary-science fields, is required to understand the human–environmental–social system.

1.1 Human–Environmental–Social System

Generally, the word "system" is defined as an entity that continuously produces a certain "output" as long as a certain "input" is provided. The input comprises raw valuables such as fuel, energy, or various resources. The output comprises valuables that are preferable to the input—i.e., those that directly improve our life on both qualitative and quantitative levels, such as processed goods, services, safety, ease, and a comfortable environment.

In 1977, W. J. Karplus, who was instrumental in establishing the new discipline of systems engineering, presented the concept of a "system rainbow".[1] As shown in Fig. 1.1, this rainbow is a spectrum for modeling and simulation, wherein all models can be placed somewhere between a "white-box model" and a "black-box model." With white-box model, one can expect to precisely predict a result because an established mathematical model can reproduce what actually happens. The mathematical model can transparently capture the inner structure of a system. Electrical systems, mechanical systems, fluid & thermal systems, and chemical systems would be classified as white-box models. Hence, any of these systems can be an objectives for control as well as design. By contrast, bio-systems, economic systems, and human social systems are likely black-box models, as one has to rely on a certain statistical model to analyze their cause–effect relations; thus, we cannot deterministically predict a coming result, but must rather statistically infer it. More specifically,

[1] Karplus (1977).

J. Tanimoto, *Evolutionary Games with Sociophysics*, Evolutionary Economics and Social Complexity Science 17, https://doi.org/10.1007/978-981-13-2769-8_1

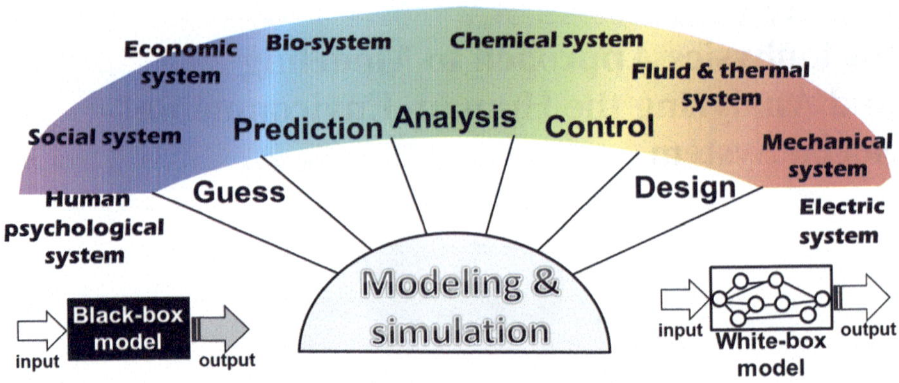

Fig. 1.1 Black-box and white-box models mapped in the System Rainbow proposed by Karplus[1]

dealing with a psychological system, we cannot help but rely on a fully black-box model that only describes an input–output relation. In such a case, the system can no longer be an "objective" for control to obtain preferable results; it simply becomes a "target" for rough estimation. According to the concept proposed by Karplus, a complex system may be expected to exhibit a black-box structure.

In fact, the real world is composed of many complex systems at various layers. The complexity in each system can be approximately quantified by measuring stochastic propensity such as probabilistic deviation and higher moments of the system. Although there are many factors affecting such stochastic deviations, human factors working through individual decision-making processes, subjective intentions, and actual behavior would be much more significant than the other probabilistic factors. Thus, it is speculated that, by explicitly modeling a "human" as a white box, it becomes extremely difficult even to model a mechanical system with a human interface or modeling a system that integrates both physical and human structures. Meanwhile, when we discuss an "environmental system" such as global weather, urban climate, or even a micro indoor environment, we are discussing what has been conventionally regarded as a physical system that is entangled with many sub-elements; in these cases, we can establish a set of governing equations by referring to physical principles such as heat-transfer theory and fluid dynamics. Such an environmental system, however, is both explicitly and implicitly influenced by human behavior, or more precisely, by the human wish to control an environmental system to obtain preferable gain (i.e., payoffs in game theory). This fact inevitably connects the environmental and human systems in interactive complex action–reaction chains. In other words, a human individual in view of a social system (i.e., a group of aggregated human individuals) and an environmental system (depicted by explicit governing equations) are mutually dependent, or interdependent. This heuristic observation suggests that human, environmental, and social systems should be simultaneously modeled in a single framework. This is why we require the concept of **human–environmental–social systems**. Any

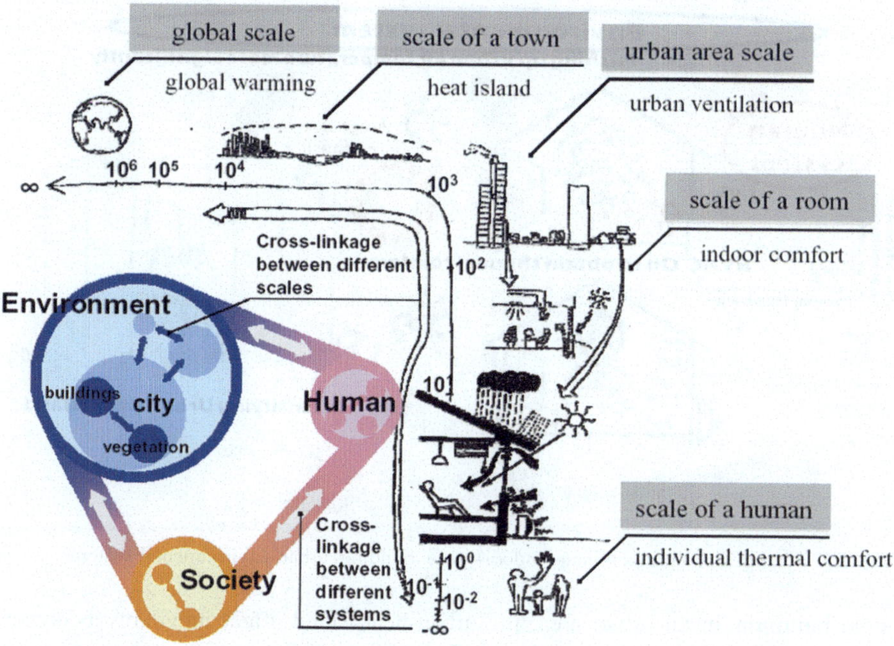

Fig. 1.2 A wide range of spatial scales over which environmental systems act, and the concept of the human–environmental–social system

model of the human–environmental–social system must consider the three spheres (the human system, the environmental system, and the social system) with mutual connecting linkages in a simultaneous plane.

Figure 1.2 demonstrates the concept of the human–environmental–social system. The three systems are respectively driven by different principles. The environmental system is driven by physical properties, such as heat (temperature is its potential), kinetic energy (velocity is its potential), and various scalars, all of which are much more objective than subjective properties such as human intention and emotion, which drive human and social systems. More importantly, the environmental system can be considered as a nested structure in terms of spatial scale, such as the structure of a city, a building, an indoor space, and a human micro-climate, as schematically explained in Fig. 1.2. One of our biggest questions is how we appropriately establish modeling of human intention and social dynamics, which should be connected with the environmental system by means of cross-linkages, as shown in the figure.

Let us refer to another example. Figure 1.3 presents a human–environmental–social system that captures why we heavily rely on cooling systems (heating, ventilating, and air-conditioning, expressed as "HVAC" in the figure) during peak summer time. If a room's air temperature is too hot, a resident would turn on the HVAC system. This behavior can be predicted by a state-transition probability, for example, by applying the theory of Markov chains, which is essentially obtained by intensive experimental efforts. As the frequency of the HVAC turn-on state increases

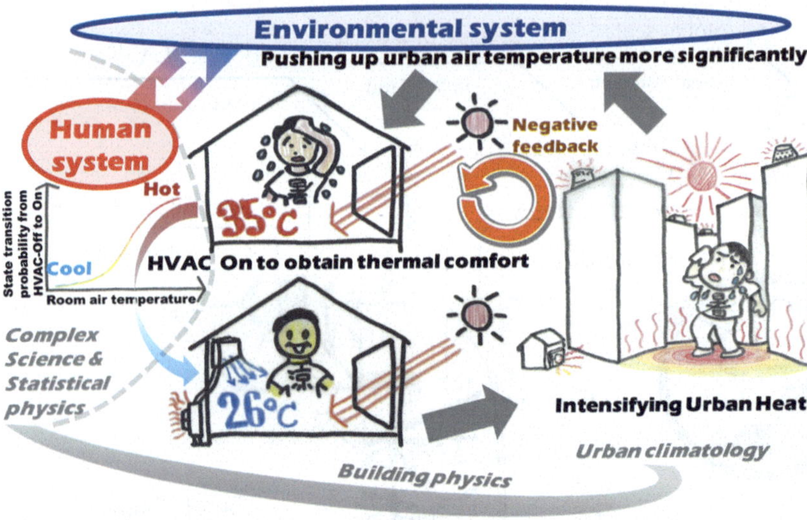

Fig. 1.3 Synergetic loop in a human–indoor-space–building–urban-environmental system

amid buildings in an urban area, the urban-heat-island effect inevitably becomes more significant due to heat exhaust. This dramatically increases the urban air temperature. Consequently, urban residents have to increasingly rely on mechanical cooling systems. This is a typical feedback loop with a negative impact upon the human–environmental system. Under the "human system" sphere, there is a state-transition-probability function that stochastically predicts how often a resident turns on and off an air-conditioning system. Other than this element, all components in an environmental system can be described by means of a white-box model approach; there is a set of physical equations drawn from building physics and urban climatology. For a resident, turning on the HVAC might be the best solution for maximizing their own gain (benefit or payoff) in terms of thermal comfort. However, this action is bad in the public sense when the vast majority of urban residents make the same choice, resulting in a mass (social) effect. Elevated urban temperature calls for more overwhelming dependence on HVAC, resulting in a large energy consumption, leading to further deterioration of the urban heat island. In this scenario, the urban air temperature, or urban environment, is a "public good". Anyone can access public space and utilize it by exhausting heat there to maintain HVAC operation in a "private" space. However, when quite a few people behave in the same manner, public goods run the risk of collapsing. This is commonly observed in various environmental problems, classically discussed as the "tragedy of the commons"[2] in the context of game theory. In sum, to entirely package this kind of story with various sub-ordinates in a single holistic model that

[2]Hardin (1968).

includes both environmental and human systems in an interactive way, the question of how we model a "human system" should be carefully considered.

1.2 Evolutionary Game and Sociophysics

Knowledge derived from **operations research** (**OR**), one of the greatest products of twentieth-century applied mathematics, may offer us some fundamental ideas for coping with the abovementioned problem. Complex science and statistical physics may help to solve this specific question. New trends in statistical physics have begun to expand to fields conventionally studied in social science as well as to ergonomics, which mathematically treats a group of human individuals as a mass of molecular particles.

Emerging from all of these efforts is a new concept with a high-profile name: "**sociophysics**". Socio physics is transdisciplinary; it arises from meetings between statistical physicists, information scientists, theoretical biologists, and various social scientists to seek possible new applications for many socially recognized practical problems and questions. One important baseline is that sociophysics fully relies upon mathematical modeling that is strongly propelled by physics concepts, thereby enabling us to quantify, analyze, and predict a result, even when there is some noise or stochastic deviation. It is important to note that the increasing power of computational technology is a key support in sociophysics. In fact, in recent years, it has been possible for anyone to easily conduct a huge numerical experiment in which complex social dynamics and the behavior of many human agents can be observed with the help of knowledge modeled using artificial intelligence (i.e., machine learning, neural networks, fuzzy theory, genetic algorithms, etc.). This is an "artificial society," which encourages many social physicists to tackle problems within a computational world. This approach is known as **multi-agent simulation** (**MAS**).

Evolutionary game theory plays one of the most important roles in terms of mathematically dealing with human intention. This book is primarily concerned with evolutionary game theory.

What do you imagine when you hear the term "game"? Is it cards? Chess? Or perhaps arcade games and video games? Certainly, these are all games. Some may allegorize a human life to a game, which seems quite plausible as an aphorism. All of these can be covered by "game theory," which is a mathematical framework for modeling the human decision-making process using some simplifications and premises (Fig. 1.4).

One of the most important problems with which game theory has been concerned for several decades is why humans cooperate altruistically as opposed to always giving in to egocentric ideas.

Why do we cooperate? Why do we observe many animals cooperating? The mysterious labyrinth surrounding the emergence of cooperative behavior in the real world has attracted much attention. The classical metaphor for investigating this

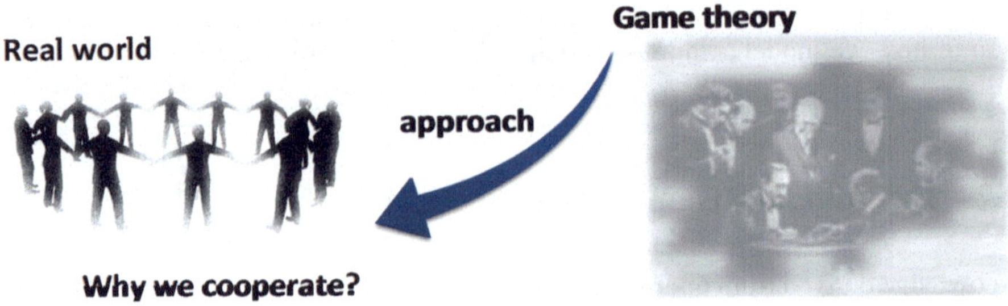

Fig. 1.4 How are humans able to establish reciprocity when encountering a social-dilemma situation in the real world?

social problem is the prisoner's dilemma (PD) game, which has been considered the most appropriate approach and is frequently used as a template for social dilemmas.

Evolutionary game theory[3] evolved from game theory by merging it with the basic concept of Darwinism to incorporate the idea of time evolution; this is partially lacking in the original game theory, which primarily deals with equilibrium.

Game theory was established in the mid-twentieth century as a novel concept by von Neumann and Morgenstern.[4] After its inception, they provided the biggest milestones in driving the theory forward; it was then made more applicable in various fields (not only economics, but also biology, information science, statistical physics, and other social sciences) by John Nash, one of the three game theorists who were awarded the Nobel Prize in Economics. He did this by formulating the concept of Nash Equilibrium.[5] Another important contribution to evolutionary game theory was provided by Maynard Smith[6] in the 1980s. He formulated a central concept of evolutionary game theory called the evolutionarily stable strategy (ESS). In the 1990s, with the rapid growth of computational capabilities, MAS became a major driver for evolutionary game theory, allowing researchers to easily construct a flexible model, free from the premises that previous theoretical frameworks presumed.[7] This enabled game players in these models to behave more intelligently and realistically. Consequently, this framework has attracted many scholars seeking answers to the question of why we can observe so much evidence for reciprocity working in real human social systems and also among animal species even in situations of severe social dilemma, where classical game theory would predict that players act defectively. As one example, the theory shows that all players would be trapped as complete defectors in the case of the PD, which will be

[3]There are many good resources concerning evolutionary game theory, such as Weibull (1995).

[4]Von Neumann and Morgenstern (1944).

[5]Nash (1949).

[6]Maynard Smith (1982).

[7]The classical game theory assumes infinite population and perfect anonymity among the players. This is called a well-mixed situation. Also, the players are presumed to act in an ideally rational way.

explained later in this book. However, we can observe a lot of evidence that opposes this finding in the real world, where we ourselves and some other animal species show social harmony with mutual cooperation in our respective social contexts.

Since these developments, thousands of papers have been produced reporting on research performed by means of computer simulations. Most follow the same pattern: in each of the new models they build, a priori is shown with numerical results indicating more enhanced cooperation than what theory predicts. These results are meaningful from the constructivist viewpoint, but are less persuasive in answering the question: "what is the substantial mechanism that causes mutual cooperation to emerge instead of defection?"

Nowak successfully made progress in understanding this problem, to some extent, with his ground-breaking research.[8] He proved theoretically that all reciprocity mechanisms that bring mutual cooperation can be classified as one of five types, and that all of them, amazingly, have similar inequality conditions for evolving cooperation because of the so-called Hamilton Rule (kin-selection rule). Nowak calls all these fundamental mechanisms "social viscosity." The Hamilton Rule[9] finally solved the puzzle originally posed by Charles Darwin in his book *The Origin of Species* (1859), of why sterile social insects, such as honey bees, leave reproduction themselves to caring their sisters by arguing that a selection benefit to related organisms would allow the evolution of a trait that confers the benefit but destroys the individual at the same time. Hamilton clearly deduced that kin selection favors cooperative behavior as long as the inclusive-fitness surge resulting from the concept of relatedness is larger than the dilemma strength. This bombshell finding by Nowak, even though he assumed several premises in his analytical procedure, demonstrates that all reciprocity mechanisms ever discussed can be explained with a simple mathematical formula very similar to the Hamilton Rule, implying that "Nature is controlled by a simple rule." The Nowak classifications—kin selection, direct reciprocity, indirect reciprocity, network reciprocity, and group selection—successfully added a new level to the controversy, but there have still been many papers that attempt to answer questions along the lines of "how much cooperation thrives if you rely on our particular model." Such questions remain relevant because Nowak's deduction has several limitations, and may therefore differ from the real reciprocity mechanism. In fact, among the five mechanisms, network reciprocity has been very well investigated, as researchers believe that complex social networks may relate to emerging mutual cooperation in a social system.

All of these developments have driven evolutionary game theory to become one of the most exciting fields in science. With great contributions from both qualitative and quantitative points of view, it seems to have reached the corner of a "phase-change," from inquiry to applications. It is therefore desirable to seek new possibilities for solving practical social problems using evolutionary game theory. This book presents one such direction.

[8]Nowak (2006).

[9]Hamilton (1964).

1.3 Structure of This Book

This book follows the structure of the previous book "Fundamentals of Evolutionary Game Theory and its Applications".[10] Recent advancements in evolutionary game theory and its rich applications will be described within. However, the present book does not try to cover all fundamental aspects of classical and evolutionary game theory; rather, we intend to share the latent strength of evolutionary game with readers who can then apply it to their own practical problems.

Chapter 2 describes the latest findings concerning archetypical games; 2-player and 2-strategy (hereafter, 2 $\times$ 2) games and multi-player games, which provide important fundamental concepts for applications.

In Chap. 3, we apply evolutionary game theory to traffic-flow analysis. Traditionally, such analysis has been quantified by applying the principles of fluid dynamics. One-dimensional fluid-pipe flow offers a good analogy, even though flowing objects, e.g., vehicles, are not continuous like water, oil, and air, but rather discrete. However, when one considers human factors, such as drivers' contributions to a flow field, the traditional concept fails to provide meaningful information. At this point, it would be a quite challenging idea to apply the evolutionary game theory to traffic-flow analysis.

In fact, a road can be regarded as a finite "resource". Each driver mutually competes to maximize their own benefit (for example, by minimizing travel time). To ensure this outcome, a rational driver may behave in egocentric ways like making frequent lane changes, even though they sometimes involve the risk of a traffic accident, inevitably invoking traffic turbulence with the emergence of a traffic jam that significantly reduces the social efficiency. This scenario, which could be described by a certain social dilemma, is illustrated in Fig. 1.5. Indeed, the road is a public good in this specific context, which any driver can access and utilize. However, as drivers, motivated by the individual payoff, start to utilize the road at the maximum level, the public good would consequently crumble leading to jams causing trouble for everyone. In a nutshell, the egocentric aspiration of "going first" jeopardizes the public benefit. One of the main motivations for us to apply evolutionary game theory is to determine what class of social dilemma hides behind a traffic flow, whether it depends upon the traffic density, and whether it relates to the so-called traffic-flow phase, e.g., free-flow or jam phases. If we can clarify the fundamental mechanism connecting the traffic flow with the social-dilemma structure, a meaningful solution to avoiding, or at least relaxing, traffic jams can be proposed.

Chapter 4 provides another example of a practical application to which evolutionary game theory has been impressively applied, known as the "vaccination game." The spread of an infectious disease like influenza is a pressing social concern. Traditional epidemiology has successfully established a powerful tool to measure how such an infectious disease diffuses in a complex human network,

[10] Tanimoto (2015).

Fig. 1.5 Traffic flow contains social dilemmas, which can be approached by evolutionary game theory

which operates on a series of models of theoretical epidemiology, effectively described by a set of ordinary differential equations (ODEs), known as the SI, SIS, SIR, and SIER models, as well as variants. Similar to the situation with traffic-flow analysis, the conventional theory may be insufficient to precisely account for the extent to which humans undergo pre-emptive vaccination to protect themselves, or whether people's vaccination efforts can cause a disease to become extinct or to break out. This situation may bring social dilemma because one can avoid costly vaccination if the majority of people around them undergo vaccination, which is termed "herd immunity." Once herd immunity is established, one can free-ride on others' effort. Thus, we can call herd immunity a public good, similar to a road.

In both Chaps. 3 and 4, we discuss a new concept for analyzing the complex human–environmental–social system by dovetailing a conventional theory, namely fluid dynamics or epidemiology, with evolutionary game theory. One feature common to both examples is that the public good would decrease if each social member primarily seeks to maximize individual benefit rather than maintain the health of the public goods. These respective systems intrinsically have a social-dilemma structure. The term dilemma, although it has already been used in the text, is rigidly defined as a situation wherein the choice that each individual makes among their possible strategies does not accord with the preferred state from the social point of view. More precisely, in terms of mathematics, it can be said that the Nash Equilibrium is not consistent with some social optimum. By applying an evolutionary-game-theoretic approach in such intriguing ways, we can underline the social-dilemma

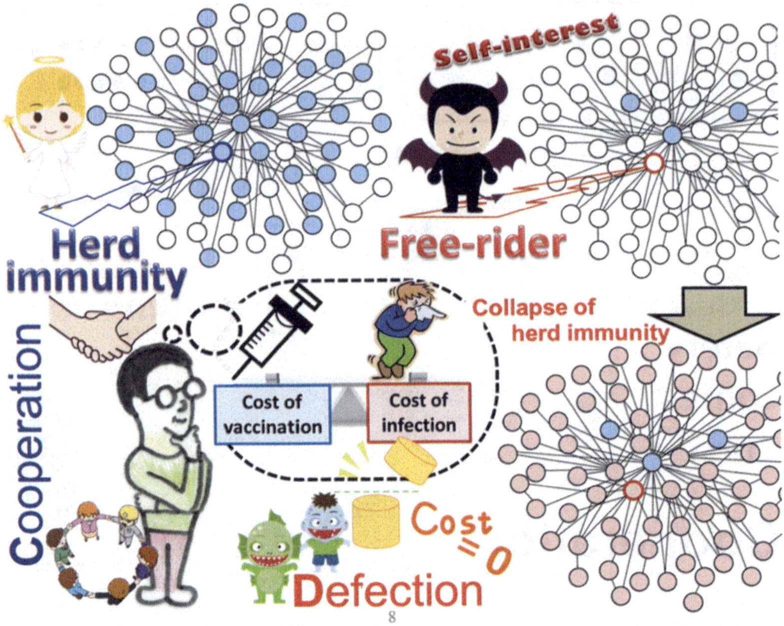

Fig. 1.6 The spread of an infectious disease and predicting people's vaccination coverage offers another example of a complex social dilemma that can be approached by evolutionary game theory

structure hidden behind surface phenomena, i.e., a traffic- flow field and the dynamic process of a disease spreading, respectively; thus, we may be able to pose a meaningful provision for breaking such dilemmas, leading all people to a socially desirable situation (Fig. 1.6).

References

Hamilton, W. D.; The genetical evolution of social behavior I and II, *Journal of Theoretical Biology* 7, 1–16 and 17–52, 1964.

Hardin, G.; The tragedy of the commons, *Science* **162**, 1243–1248, 1968.

Karplus, W. J.; The spectrum of mathematical modeling and system simulation, *Mathematics and Computers in Simulation* **19 (1)**, 3–10, 1977.

Maynard Smith, J.; Evolution and the Theory of Games. Cambridge University Press (Cambridge), 1982.

Nash, J. F.; Equilibrium points in n-person games, *Proceedings of the National Academy of Science of the United States of America* 36 (1), 48–49, 1949.

Nowak, M. A.; Five rules for the evolution of cooperation, *Science* 314, 1560–1563, 2006.

Tanimoto, J.; Fundamentals of Evolutionary Game Theory and its Applications, Springer (Tokyo), 2015.

Von Neumann, J., Morgenstern, O.; Theory of Games and Economic Behavior, Princeton University Press (Princeton), 1944.

Weibull, J. W.; Evolutionary Game Theory, MIT Press (Cambridge), 1995.

Chapter 2
Evolutionary Game Theory

This chapter presents the fundamental concept of evolutionary game theory. Originally, game theory referred to a mathematical framework for the human decision-making process, containing various variants—whether a game is zero-sum (constant-sum), meaning that if one player wins the other must lose, or non-zero-sum (non-constant-sum); whether a game is symmetric, with both a focal player and an opponent sharing a common payoff structure, or asymmetric; whether a game is 2-player or multi-player; whether a game has two strategies or multiple—and so forth. In any case, classical game theory primarily concerns determining game equilibrium, or a game solution, which can be understood as a steady-state solution or a static solution in the field of conventional science and engineering. On the other hand, evolutionary game theory is rather concerned with the time-evolution of a system. This theory, as well as profound observation of the dynamical process, may allow us to solve some scientific questions—e.g., why cooperation is commonly observed in many animal species, including human beings.

To begin, we examine the appropriate treatment of 2-player and 2-strategy (2 × 2) games, which constitute the most important archetype among evolutionary games. Subsequently, multi-player and 2-strategy games are also introduced. In the latter parts of this chapter, we present one of the latest findings concerning how multiple reciprocity mechanisms may relax a social dilemma.

2.1 2-Player & 2-Strategy Games

2.1.1 Fundamental Framework

In this section, the 2-player–2-strategy game (abbreviated two-by-two game or 2 × 2 game) is explained. It is the most important and fundamental template for understanding evolutionary games.

J. Tanimoto, *Evolutionary Games with Sociophysics*, Evolutionary Economics and Social Complexity Science 17, https://doi.org/10.1007/978-981-13-2769-8_2

Let us suppose that we have an unlimited population. The individuals (hereafter, sometimes called "agents") are well-mixed, e.g., there is no spatial special structure mutually connecting individuals, no iteration of games, no mechanisms to specify individuals such as tags, skin color, or mutual communication to identify one another and add "social viscosity" (Nowak[1]) to the system. This specific situation is called an "infinite & well-mixed population", which implies that each agent never plays the same opponent again.

From the infinite & well-mixed population, two individuals are selected at random and made to play the game. The game uses two discrete strategies (as shown in Fig. 2.1): cooperation (C) and defection (D). The pair of players receive payoffs in each of the four combinations of C and D. A symmetrical structure is assumed between the two players. In Fig. 2.1, the payoff of player 1 (the "row" player) is represented by the entries preceding the commas; the payoff of player 2 (the "column" player) by the entries after the commas. The payoff matrix is denoted by $\begin{bmatrix} R & S \\ T & P \end{bmatrix}$. Depending on the relative magnitudes of the matrix elements P, R, S, and T, the game can be divided into 4 classes—the Trivial game with no dilemma; the Prisoner's Dilemma (sometimes abbreviated PD); Chicken (also known as the Snow Drift or Hawk–Dove Game); and Stag Hunt (sometimes abbreviated SH). It is quite natural that the system described by the payoff matrix has 4 degrees-of-freedom, since the number of variables is four.

To expand our standpoint beyond classical game theory to evolutionary game theory, let us here introduce the concept of "time".

The strategies (called "hands" in some contexts) adopted by an agent are cooperation (C) or defection (D), expressed by the following state vectors[2]:

$$\text{Strategy C; } {}^{T}\mathbf{e1} = (1 \quad 0), \tag{2.1a}$$

$$\text{Strategy D; } {}^{T}\mathbf{e2} = (0 \quad 1). \tag{2.1b}$$

The payoff matrix of the game structure is

Fig. 2.1 Payoff matrix of 2 × 2 game

Agent 2 / Agent 1	Cooperation (C)	Defection (D)
Cooperation (C)	R, R	S, T
Defection (D)	T, S	P, P

[1]Nowak (2006).

[2]To learn the details of the following mathematical procedure, please see Tanimoto (2014, 2015).

$$\begin{bmatrix} R & S \\ T & P \end{bmatrix} \equiv \mathbf{M} \tag{2.2}$$

Moreover, the proportions of agents adopting strategies C and D at a given time (referred to as the strategy ratio) are defined by s1 and s2 respectively. These strategy ratios are expressed as;

$${}^{T}\mathbf{s} = (\, s_1 \quad s_2 \,). \tag{2.3}$$

From the simplex condition, we obtain

$$s_2 = 1 - s_1. \tag{2.4}$$

The validity of Eqs. (2.2 and 2.3) should be understood from the following vector-matrix equation describing a contest between two agents adopting strategy D, in which the outcome is P:

$$\pi_{DD} = (\,0 \quad 1\,) \cdot \begin{bmatrix} P & S \\ T & P \end{bmatrix} \begin{pmatrix} 0 \\ 1 \end{pmatrix} = P. \tag{2.5}$$

A variant form of Eq. (2.5) can also compute the payoff when one strategy plays a game **M** against another with a different strategy. The expected payoff when an agent using strategy C battles with a randomly sampled agent at the present time, expressed as strategy ratio s, is

$${}^{T}\mathbf{e1} \cdot \mathbf{M}\,\mathbf{s}.$$

Similarly, the expected payoff when an agent using strategy D fights a randomly sampled agent at the present time, expressed as strategy ratio s, is

$${}^{T}\mathbf{e2} \cdot \mathbf{M}\,\mathbf{s}.$$

The replicator dynamics are defined as the strategy-ratio dynamics of strategy i, expressed as

$$\frac{\dot{s}_i}{s_i} = {}^{T}\mathbf{ei} \cdot \mathbf{M}\,\mathbf{s} - {}^{T}\mathbf{s} \cdot \mathbf{M}\,\mathbf{s}. \tag{2.6}$$

The dimensionless quantity on the left-hand side of (2.6), obtained by dividing $\dot{s}_i$ by the strategy ratio itself, is called the coefficient of deviation in the field of statistics; it indicates the extent of strategy change. This quantity is determined as the difference between the payoff for strategy i playing against the social average at a given time and the expected social payoff at that time. As a driving rule letting a certain dynamical system evolve in time, replicator dynamics seem appropriate for the following reason: after a game, the successful strategies (those achieving higher payoffs than the average accumulated by the strategy ratio) will increase at the next time step, whereas less successful strategies will decrease. The ratio of this extent is thought to be decided by comparison with the aforementioned level of "success." In such a system, good conduct is rewarded whereas bad conduct is punished (a form of survival of the fittest). Selection mechanisms in the natural world (including human social systems) tend to operate in this manner.

Substituting Eqs. (2.1), (2.2), and (2.3) into Eq. (2.6) and explicitly writing the elements, we obtain

$$\begin{cases} \dot{s}_1 = [(R-T)\cdot s_1 - (P-S)\cdot s_2]\cdot s_1\cdot s_2 \\ \dot{s}_2 = -[(R-T)\cdot s_1 - (P-S)\cdot s_2]\cdot s_1\cdot s_2 \end{cases}. \tag{2.7}$$

Note that when the right-hand side of (2.7) = 0, the equation becomes cubic in s_l and s_2; that is, the system contains three equilibrium points. Two of these are self-evident:

$$(\, s_1 \quad s_2\,) = (\,1 \quad 0\,) \equiv \mathrm{s}^* \mid \text{C-dominate}, \tag{2.8a}$$

$$(\, s_1 \quad s_2\,) = (\,0 \quad 1\,) \equiv \mathrm{s}^* \mid \text{D-dominate}. \tag{2.8b}$$

At the former point, all individuals ultimately become cooperative (all-cooperators state); the latter leads to the defection state (all-defectors state), implying C-dominance and D-dominance, respectively. The remaining equilibrium point is obtained by simultaneously solving Eq. (2.7), setting [. . .] on the right-hand side to 0 and eliminating s_2 through Eq. (2.4):

$$(\, s_1 \quad s_2\,) = \left(\frac{P-S}{P-T-S+R} \quad \frac{R-T}{P-T-S+R} \right) \equiv \mathrm{s}^* \mid \text{Polymorphic}. \tag{2.8c}$$

This third equilibrium point lies within [0, 1], depending on the values of *P*,*R*,*S*, and *T*. In this case, the dynamics become polymorphic or bi-stable. Equation (2.8c) defines an internal-equilibrium point.

Once the three equilibrium points are obtained, the signs of the eigenvalues of the Jacobian matrix at each equilibrium point are determined, and the equilibrium points are assessed as sink, source, or saddle.

To this end, we re-write Eq. (2.7) as follows:

$$\dot{s}_1 \equiv f_1(s_1, s_2), \tag{2.9a}$$

$$\dot{s}_2 \equiv f_2(s_1, s_2). \tag{2.9b}$$

From Eq. (2.4), we observe that $f_1 = -f_2$.

Meanwhile, let us consider an arbitrary continuous dynamical system in which the system's state equations are expressed by a non-linear function **f**:

$$\frac{d\mathbf{x}}{dt} = \dot{\mathbf{x}} = \mathbf{f}(\mathbf{x}). \tag{2.10}$$

In general, non-linear functions are approximated as linear over infinitesimal intervals by Taylor expansion. Expanding the right-hand side of Eq. (2.10), we obtain;

$$\begin{aligned} &\mathbf{f}(\mathbf{x}) = \mathbf{f}\,(\mathbf{x}^*) + \mathbf{f}'(\mathbf{x}^*)(\mathbf{x} - \mathbf{x}^*) + \frac{\mathbf{f}''(\mathbf{x}^*)}{2!}(\mathbf{x} - \mathbf{x}^*)^2 + \cdots \\ &\Leftrightarrow \mathbf{f}(\mathbf{x}) \cong \mathbf{f}\,(\mathbf{x}^*) + \mathbf{f}'(\mathbf{x}^*)(\mathbf{x} - \mathbf{x}^*). \end{aligned} \tag{2.11}$$

From the definition of an equilibrium point, note that $\mathbf{f}\,(\mathbf{x}^*) = \mathbf{0}$ (this should be evident by substituting $\frac{d\mathbf{x}}{dt}\big|_{\mathbf{x}=\mathbf{x}*} = \mathbf{0}$ into Eq. (2.10)). By ignoring higher-derivation terms, Eq. (2.11) is approximately equal to;

$$\mathbf{f}(\mathbf{x}) \cong \mathbf{f}'(\mathbf{x}^*)(\mathbf{x} - \mathbf{x}^*). \tag{2.12}$$

Equation (2.12) can be approximated as a linear equation as follows:

$$\mathbf{f}(\mathbf{x}) = \mathbf{f}'(\mathbf{x}^*)(\mathbf{x} - \mathbf{x}^*) = \mathbf{f}'(\mathbf{x}^*)\mathbf{x} - \mathbf{f}'(\mathbf{x}^*)\mathbf{x}^*. \tag{2.13}$$

The first term on the right of (2.13) is first-order in x, while the second term is constant. Backing to the mathematical common-sense that **A** is called the transition matrix in case of a liner dynamical system; $\dot{\mathbf{x}} = \mathbf{A}\mathbf{x}$. The original non-linear system (2.10) has been simplified to the linear system (2.13). Thus, the transition matrix is $\mathbf{f}'(\mathbf{x}^*)$. According to the theory of a linear-dynamical system, we can say whether each equilibrium point absorbs or expels a specific time-evolutionary episode by identifying whether the slope of the system at each equilibrium point is positive or negative, which can be evaluated by the signs of the eigenvalues corresponding to the equilibrium points of this matrix.

The transition matrix happens to be the Jacobian matrix of tangent gradients of the multi-variable vector function:

$$\mathbf{f}'(\mathbf{x}^*) = \frac{\partial\,\mathbf{f}(\mathbf{x})}{\partial\mathbf{x}}\bigg|_{\mathbf{x}=\mathbf{x}*} = \begin{bmatrix} \frac{\partial f_1(\mathbf{x})}{\partial x_1} & \cdots & \frac{\partial f_1(\mathbf{x})}{\partial x_n} \\ \vdots & \ddots & \vdots \\ \frac{\partial f_n(\mathbf{x})}{\partial x_1} & \cdots & \frac{\partial f_n(\mathbf{x})}{\partial x_n} \end{bmatrix}_{\mathbf{x}=\mathbf{x}*}. \tag{2.14}$$

Here, let us apply Eq. (2.14) to equations up to (2.9). Each element of the Jacobian of Eq. (2.9) is calculated as;

$$\begin{cases} \dfrac{\partial f_1}{\partial s_1} = -\dfrac{\partial f_2}{\partial s_1} = 3(-R+S+T-P){s_1}^2 & \text{(a)} \\ \quad +2(R-2S-T+2P)s_1+S-P, & \\ \dfrac{\partial f_1}{\partial s_2} = -\dfrac{\partial f_2}{\partial s_2} = -3(-R+S+T-P){s_1}^2 & \text{(b)} \\ \quad -2(R-2S-T+2P)s_1-S+P. & \end{cases} \tag{2.15}$$

The Jacobian matrix $\mathbf{J} = \begin{bmatrix} \frac{\partial f_1}{\partial s_1} & \frac{\partial f_1}{\partial s_2} \\ \frac{\partial f_2}{\partial s_1} & \frac{\partial f_2}{\partial s_2} \end{bmatrix} = \begin{bmatrix} \frac{\partial f_1}{\partial s_1} & \frac{\partial f_1}{\partial s_2} \\ -\frac{\partial f_1}{\partial s_1} & -\frac{\partial f_1}{\partial s_2} \end{bmatrix}$ is a 2×2 matrix, so its eigenvalues are easily obtained: 0 and $\frac{\partial f_1}{\partial s_1} - \frac{\partial f_1}{\partial s_2}$. Since 0 is unsigned, we need only obtain the sign of $\frac{\partial f_1}{\partial s_1} - \frac{\partial f_1}{\partial s_2}$ to establish the equilibrium conditions. Explicitly, these eigenvalues are

$$\lambda = \frac{\partial f_1}{\partial s_1} - \frac{\partial f_1}{\partial s_2} = 6(-R+S+T-P){s_1}^2 \\ +4(R-2S-T+2P)s_1 + 2(S-P). \tag{2.16}$$

1. The necessary and sufficient condition for the equilibrium point $\mathbf{s}^*|_{\text{C-dominate}}$ to be a sink is $\lambda < 0$ when substituting $(s_1 \quad s_2) = (1 \quad 0)$ into Eq. (2.16). The following condition is sought:

$$T - R = D_g < 0. \tag{2.17}$$

2. The necessary and sufficient condition for the equilibrium point $\mathbf{s}^*|_{\text{D-dominate}}$ to be a sink is $\lambda < 0$ when substituting $(s_1 \quad s_2) = (0 \quad 1)$ into Eq. (2.16). We now require that

$$P - S = D_r > 0. \tag{2.18}$$

3. The necessary and sufficient condition for the equilibrium point $\mathbf{s}^*|_{\text{Polymorphic}}$ to be a sink is $\lambda < 0$, with $(s_1 \quad s_2) = \left(\dfrac{P-S}{P-T-S+R} \quad \dfrac{R-T}{P-T-S+R} \right)$ substituted into Eq. (2.16). Noting that $\lambda = 2\frac{(R-T)(P-S)}{R-S-T+P}$, we seek the following conditions:

$$P < S \wedge R < T \Leftrightarrow P - S = D_r < 0 \wedge T - R = D_g > 0. \tag{2.19}$$

Here, let us define the gamble-intending dilemma (hereafter referred to as the GID, or the Chicken-type dilemma), D_g, and the risk-averting dilemma (hereafter referred to as RAD, or the Stag Hunt (SH)-type dilemma); D_r, respectively[3]:

$$\begin{aligned} D_g &\equiv T - R \\ D_r &\equiv P - S. \end{aligned} \tag{2.20}$$

The above conditions are summarized in Table 2.1, with the following substitution:

$$\mathbf{s}*|_{\text{Polymorphic}} = \left(\frac{P-S}{P-T-S+R} \quad \frac{R-T}{P-T-S+R} \right) = \left(\frac{D_r}{D_g - D_r} \quad \frac{-D_g}{D_r - D_g} \right).$$

Defining D_g and D_r in Eq. (2.20), the four game classes were established as PD, Chicken, SH, and Trivial, which can be summarized in Table 2.2. Here, these divisions are represented by the difference between the signs of the three equilibrium points. More importantly, which class—PD, Chicken, SH, or Trivial—is embedded in a given 2 × 2 game can be fully evaluated by both signs of D_g and D_r. In other words, the signs of D_g and D_r strictly regulate the game class for any arbitrary 2 × 2 game.

In PD, $\mathbf{s}^*|_{\text{C-dominate}}$ and $\mathbf{s}^*|_{\text{D-dominate}}$ are a source and a sink, respectively; hence, regardless of the initial cooperation proportion in [0, 1] the ultimate state is one of complete defection as $t \to \infty$.

In Chicken, $\mathbf{s}^*|_{\text{C-dominate}}$ and $\mathbf{s}^*|_{\text{D-dominate}}$ are both sources. In this case $\mathbf{s}^*|_{\text{Polymorphic}}$ (value in [0, 1]) is a sink, so regardless of the initial cooperation proportion, as $t \to \infty$, the system settles into an internal equilibrium point $\mathbf{s}^*|_{\text{Polymorphic}}$. As previously mentioned, this state does not imply that specific agents are fixed into cooperation or defection strategies, but that when the infinitely large group is viewed as a whole, the proportions of cooperation and defection players are (dynamically) steady.

In SH, the internal-equilibrium point $\mathbf{s}^*|_{\text{Polymorphic}}$ is a source, while $\mathbf{s}^*|_{\text{C-dominate}}$ and $\mathbf{s}^*|_{\text{D-dominate}}$ are both sinks. Therefore, if the initial proportion of cooperative players is smaller (or larger) than $\mathbf{s}^*|_{\text{Polymorphic}}$, the ultimate state is pure defection, (or pure cooperation) and the system is bi-stable.

In Trivial, $\mathbf{s}^*|_{\text{C-dominate}}$ is a sink and $\mathbf{s}^*|_{\text{D-dominate}}$ is a source, so regardless of the initial cooperation proportions, the pure-cooperation state is inevitable. Thus, Trivial is a game with no dilemmas.

The above discussion is summarized schematically in Fig. 2.2.

As long as an infinite & well-mixed population is assumed, the theory correctly predicts the dynamics of any symmetric 2 × 2 game that we have discussed so far. Thus, in a social dilemma modeled by a PD, there would be no way for cooperation

[3]To precisely know about GID & RAD andD_g and D_r, you should consult with; Tanimoto and Sagara (2007).

Table 2.1 2 × 2 game dynamics derived analytically

Game class	Phase	Nash equilibrium	Sign of D_g	Sign of D_r	Each point sink, source, or saddle		
					(1,0)	(0,1)	$\left(\frac{D_r}{D_g - D_r} \quad \frac{-D_g}{D_r - D_g}\right)$
PD	**D**-Dominate	(0,1)	+	+	Source	Sink	Saddle
Chicken	Polymorphic	$\left(\frac{D_r}{D_g - D_r} \quad \frac{-D_g}{D_r - D_g}\right)$	+	–	Source	Source	Sink
SH	Bi-stable	(0,1) or (1,0)	–	+	Sink	Sink	Source
Trivial	**C**-Dominate	(1,0)	–	–	Sink	Source	Saddle

Table 2.2 Class type in a 2 × 2 game

Game class	GID? (Dg > 0?)	RAD? (Dr > 0?)	Dilemma as a whole?
Prisoner's Dilemma; PD	Yes	Yes	Yes
Chicken(Snow Drift; Hawk-Dove)	Yes	No	Yes
Stag Hunt; SH	No	Yes	Yes
Trivial	No	No	No

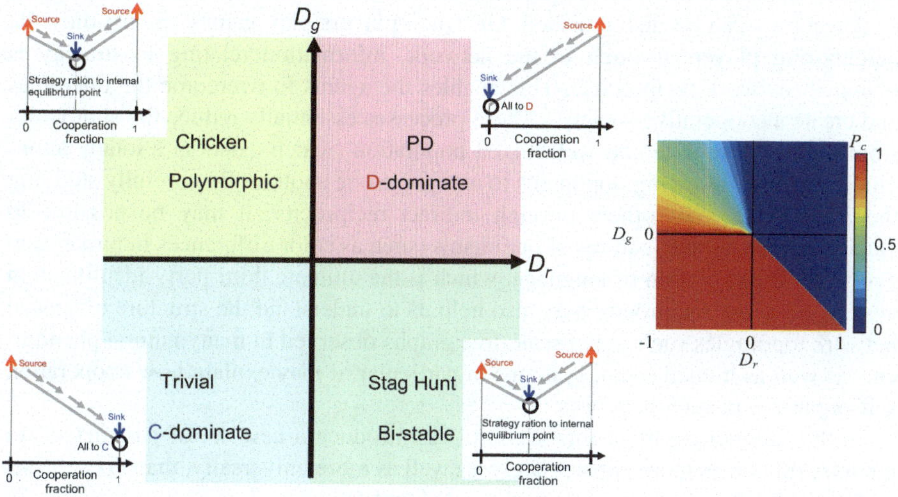

Fig. 2.2 Phase diagram of dynamics classified by D_g and D_r for a two-by-two game and a summary of the dynamics of each game class (left panel). Cooperation fraction at equilibrium when an infinite and well-mixed population with replicator dynamics is assumed presuming an initial P_c of 0.5. PD and Trivial are colored blue and red, respectively, since D-dominated and C-dominated phases are established. In the Chicken game region, a gradual shift of the cooperation fraction at equilibrium is observed due to the polymorphic phase. In the SH-game region, bi-stable shows twofold phases, either absorbed all cooperation or all defection

to dominate over defection. In the natural world, however, cooperative behavior is ubiquitous—not only in human societies, but also among social insects such as ants and bees. One question coming to our mind is what mechanisms need to be added to the original situation—i.e., an infinite & well-mixed population—to promote cooperation among agents?

Over the last decades, the puzzle of what can be called the "supplementary framework" for solving a dilemma has been of heavy concern to biologists, physicists, mathematicians, and information scientists. Many studies have taken a simulation approach and reported that their model, which introduces a new, specific additional framework, somehow enables enhanced cooperation compared to the default presuming an infinite & well-mixed population. However, these studies not directly answered the above question.

Among these, Nowak[1] showed there are the five fundamental protocols for mitigating or canceling dilemmas,[4] summarized as in Fig. 2.3. The mechanisms of these protocols are governed by very ordinary and beautiful mathematical expressions similar to those of kin selection.[5] Nowak refers to these mechanisms as "**Social Viscosity**." Under a circumstance featuring a well-mixed population, each game is played by a single person whose next encounter is unknown (well-mixed). But, in repeated games between a pair of individuals (direct reciprocity),[6] or observing the tag of an opponent (indirect reciprocity), the behavior of an opponent—cooperation or defection—can be distinguished. Or, when players play games against only the neighboring players throughout the network, information relating to strategy is obtained (network reciprocity). This enables the agents to overcome the dilemmas and create a cooperative society.[7] These processes essentially reduce the anonymity from that of an infinite and well-mixed population (which exists in a totally anonymous state) by allowing opponents to recognize one another. By carefully studying the authentication of others through indirect reciprocity, it may be possible to elucidate how notable features of organisms (such as color differences in bird crests) evolve, or the evolution of language, which is the ultimate third-party-identification system. Network reciprocity may also help us to understand the structure of special network topologies such as the scale-free graphs observed in many natural phenomena, as well as human social systems; in particular, it may explain how cooperation self-organizes in such networks.

In the following part of this section, we introduce a new set of parameters[8] to quantify dilemmas more appropriately, as well as more universally, than does the set of D_g and D_r. These new parameters are D_g' and D_r'.

As we have discussed before, as long as an infinite and well-mixed population is presumed with replicator dynamics, an evolutionary trail can be stipulated strictly by what Table 2.1 shows. In a nutshell, whenever both D_g and D_r are fixed, the evolutionary dynamics are totally determined. In this sense, D_g and D_r are scaling parameters of dilemma strength, and the dynamics of equilibrium are determined by their values.

However, D_g and D_r are insufficient for indicating the dilemma strength when a certain specific reciprocity mechanism is introduced into a game. For example, in Fig. 2.4, we show the equilibrium-cooperation fractions of spatial-PD games on a lattice network with degree $k = 8$, in which network reciprocity, one of the Nowak's five protocols, is applied as an additional framework to the original setting, namely

[4]Strictly speaking PD satisfying $D_g = D_r$.

[5]Hamilton (1963).

[6]This situation accords with common sense. If a game is played against the same partner each time, rather than against an unknown one, both individuals should accept the cooperation option to avoid strategies leading merely to short-term profit. If both individuals take the defection option P, neither will benefit in the long term. Our daily behavior follows the former pattern.

[7]Many of these dynamics can be verified by simulation. Games are repeated between multiple agents in a simulated society; this approach is known as multi-agent simulation.

[8]Wang et al. (2015a).

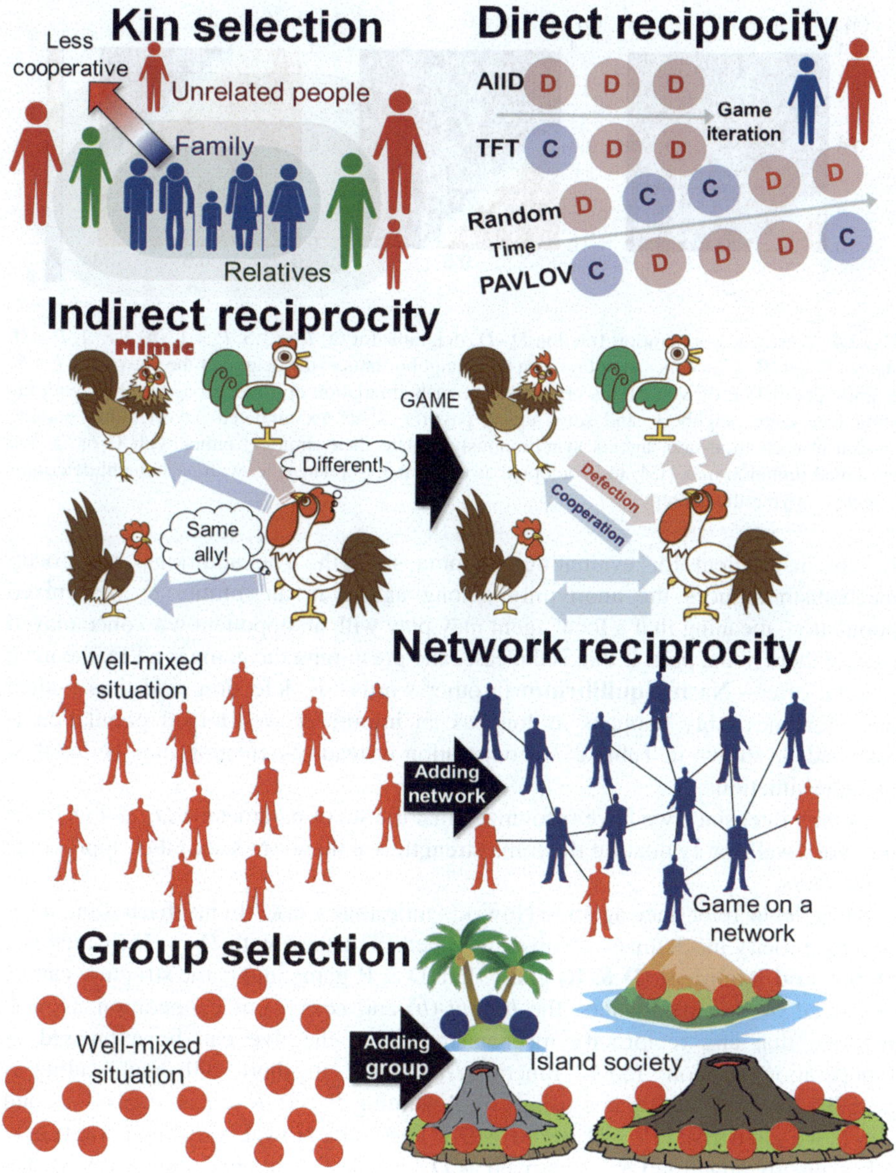

Fig. 2.3 Five basic mechanisms for dilemma resolution and an example of network reciprocity

the infinite & well-mixed population. Although these three games have the same D_g and D_r, the cooperation fractions in Fig. 2.4 differ completely depending on the value of $R - P$. The larger $R - P$ becomes, the higher the equilibrium-cooperation fraction. As this example evidently reveals, in an evolutionary game where a specific reciprocity mechanism is added to the default game setting, the parameters D_g and

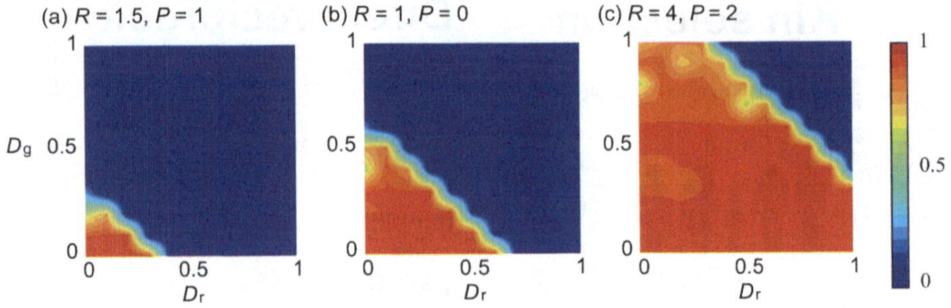

Fig. 2.4 Averaged-cooperation-fraction D_r–D_g diagrams for (**a**) $R = 1.5$, $P = 1$, (**b**) $R = 1$, $P = 0$, and (**c**) $R = 4$, $P = 2$. Games are played on an 8-neighbor lattice; the degree of the network is $k = 8$. A finite population of $N = 10^4$ is presumed. In each simulation episode, an agent plays with his immediate eight neighbors, and sums up all payoffs as his accumulated payoff. After gaming session at each time-step, agents synchronously update their strategy; either with C or D. We presumed imitation max (IM) to be adopted as the strategy-update rule, with each agent deterministically copying the strategy

D_r are insufficient for evaluating dilemma strength. The additional reciprocity mechanism reduces the anonymity among agents in an infinite & well-mixed population, meaning that a focal agent may play with an opponent who once played against them in the past. If that is the case, and presuming the game is a PD (the most rational offer—**Nash Equilibrium** in other words—is defection, and thus mutual defection inevitably happens as long as an infinite & well-mixed population is assumed, as shown in Table 2.2), cooperation instead of defection may be justified in some situations.

At any rate, since we have recognized that the set of parameters, D_g and D_r, does not work well for evaluating dilemma strength; we therefore seek other appropriate parameters.

Here, let us refer once again to Nowak's milestone work.[1] In his discussion of the five reciprocity mechanisms, Nowak assumed PD games with $D_g = D_r$, which is a **Donor and Recipient (D & R) game**. In a D & R game, the game structure can be described by two parameters, the *benefit* (b) and *cost* (c) of cooperation. Nowak reported that any reciprocity mechanism among the five can be expressed as (cooperation fraction; P_c) = function(c/b). Thus, in short, universal scaling is possible for the D & R game using c/b. Assuming $P = 0$, $R = b - c$, $S = -c$, and $T = b$, we can derive $c/b = D_g/(R - P + D_r)$. Therefore, Nowak's scaling parameter c/b has already quantified $R - P$, as well as D_g and D_r. Despite his work implying that c/b can work as a scaling parameter to express dilemma strength, Nowak's discussion is restricted to D & R games.

Inspired by Nowak's scaling concept, Tanimoto[9] proposed a set of universal scaling parameters defined as b/c_c and b/c_d. This concept can extend Nowak's

[9]Tanimoto (2009).

scaling to the general PD game by additionally considering two parameters—c_c and c_d—which imply the *focal player's costs* when his opponent offers C and D, respectively. However, Tanimoto's report assumed only PD games, and he did not demonstrate its theoretical plausibility.

We show, as below, a new set of universal scaling parameters that can be applied to all 2 × 2 games, i.e., from those with an infinite well-mixed population, which is the premise for the replicator dynamics, to those having a population with any of the five reciprocity mechanisms.

2.1.2 Concept of the Universal Scaling for Dilemma Strength

We consider 2 × 2 games composed of N agents. Let us denote the payoff matrix as;

$$\mathbf{A} \equiv [a_{ij}] = \begin{matrix} & C & D \\ C & R & S \\ D & T & P \end{matrix}. \tag{2.21}$$

When we consider an infinite well-mixed population ($N \rightarrow \infty$) and denote $x_i(t)$ as the frequency of strategy i at time t, the expected payoff of strategy i is given by $f_i = \sum_{j=1}^{2} x_j a_{ij}$. Hence, the average payoff is given by $\varphi = \sum_{i=1}^{2} x_i f_i$. The replicator dynamics can be written as.

$$\dot{x}_i = x_i(f_i - \phi). \tag{2.22}$$

From the definition of the payoff matrix, Eq. (2.21), we note $i = 1$ and $i = 2$, representing strategies C and D, respectively. Since the simplex condition $x_1 + x_2 = 1$ is kept for any time, we can reduce the two variables by introducing $x_1 = x$. Recalling (2.8), we can deduce the three equilibrium x^*-values for Eq. (2.22) by

$$x^* = 0, \quad 1, \quad \frac{P - S}{R - S - T + P}. \tag{2.23}$$

It is worthwhile to note again that in a finite well-mixed population or a population with any reciprocity mechanism, the equilibrium under replicator dynamics can no longer be given by Eq. (2.23).

According to Eq. (2.20), the payoff matrix; Eq. (2.21), can be re-written as

$$\mathbf{A} = [a_{ij}] = \begin{array}{c} \\ C \\ D \end{array} \begin{array}{c} \begin{array}{cc} C & D \end{array} \\ \begin{pmatrix} R & P - D_{\mathrm{r}} \\ R + D_{\mathrm{g}} & P \end{pmatrix} \end{array}. \tag{2.24}$$

We also note that the third equilibrium of Eq. (2.23), the so-called internal equilibrium, can be given as follows:

$$x^* = \frac{D_{\mathrm{r}}}{D_{\mathrm{r}} - D_{\mathrm{g}}}. \tag{2.25}$$

Here, let us introduce a new set of scaling parameters considering a finite population with any reciprocity mechanism by defining a new set of GIDs and RAD as D_{g}' and D_{r}', respectively:

$$D_{\mathrm{g}}' = \frac{T - R}{R - P} = \frac{D_{\mathrm{g}}}{R - P}, D_{\mathrm{r}}' = \frac{P - S}{R - P} = \frac{D_{\mathrm{r}}}{R - P}. \tag{2.26}$$

This is what we call the new universal scaling for dilemma strength. Based on this definition, the payoff matrix can be re-written again in the following form:

$$\mathbf{A} = [a_{ij}] = \begin{array}{c} \\ C \\ D \end{array} \begin{array}{c} \begin{array}{cc} C & D \end{array} \\ \begin{pmatrix} R & P - (R - P)D_{\mathrm{r}}' \\ R + (R - P)D_{\mathrm{g}}' & P \end{pmatrix} \end{array}. \tag{2.27}$$

Taylor and Nowak[10] successfully deduced that any of Nowak's five reciprocity mechanisms can be expressed by each transformation that is applied to the original 2×2 game payoff, Eq. (2.21). This allows us to derive the equilibria of each mechanism when applying the replicator dynamics on the basis of the transformed matrix.

Following their analytical approach, we below determine whether D_{g}' and D_{r}' are theoretically consistent as scaling parameters for evaluating each of the five reciprocity mechanisms. After confirming this theoretical consistency, we demonstrate that the set of these new scaling parameters works well in a finite and well-mixed population.

2.1.2.1 Direct Reciprocity

In repeated games by a pair composed of the same two agents, or when the same agents play a different game in another round, direct reciprocity stimulates cooperation.[11,12] In each round, the two agents must choose either cooperation or defection.

---

[10] Taylor and Nowak (2007).

[11] Trivers (1971).

[12] Trivers (1985).

There is a probability w that the same two agents will play another round. We assume that defectors, denoted by D, always choose defection and cooperators, C, play tit-for-tat (TFT): they start with cooperation and then follow what the other player has done in the previous move.

2.1.2.2 Indirect Reciprocity

Indirect reciprocity is based on reputation.[13,14] Unlike direct reciprocity, where the focal player's decision is based on whether their opponent has offered cooperation or defection in their previous encounter; under indirect reciprocity, the focal agent's decision is determined based on whether their opponents has offered cooperation or defection to another agent in the previous round. In fact, the focal agent chooses their strategy (offering C or D) based on their opponent's reputation, called Image Score (IS). The parameter q denotes the probability of knowing the IS of another individual, in short, of knowing whether another individual is a cooperator or a defector. Let us assume that a defector, D, always defects, whereas a cooperator, C, only defects when they know their opponent is a defector and cooperates otherwise. Thus, C cooperates with D with probability $1 - q$.

2.1.2.3 Kin Selection

The concept of kin selection arose from the idea that evolutionary games are often played between individuals who are genetically related.[15] Consider a population in which the average relatedness between interacting individuals is given by r, which is a real number between 0 and 1. In such a population, the opponent's contribution, which equals the r of the opponent's payoff, can be joined to the focal agent's payoff.

2.1.2.4 Group Selection

Group selection is based on the idea that competition occurs not only between individuals, but also between groups.[16] Here, we use the approach described by Traulsen and Nowak,[17] by which a population is subdivided into m groups. The

[13] Alexander (1987).

[14] Nowak and Sigmund (1998).

[15] Hamilton (1964).

[16] There have been many studies related to this point. Owing to space limitations, we cite only five; Wynne-Edwards (1962), Williams (1996), Wilson (1975), Maynard Smith (1976), Slatkin and Wade (1978).

[17] Traulsen and Nowak (2006).

maximum group size is n. Individuals interact with others in the same group according to a 2×2 game. The fitness of an individual is given by $1 - \omega - \omega F$, where F is the payoff and$\omega$is the intensity of selection. At each round, one individual from the entire population is chosen for reproduction proportional to fitness. The offspring is added to the same group. If the group reaches the maximum size, it can split into two groups with a probability p. Simultaneously, a randomly selected group dies to prevent the population from exploding. The maximum population size is defined as mn. With a probability of $1 - p$, however, the group does not divide, but a random individual of that group is chosen to die. We assume weak selection ($\omega < < 1$) and rare group splitting ($p < < 1$), as well as large n and m.

2.1.2.5 Network Reciprocity

Network reciprocity relies on two effects: (1) limiting the number of game opponents (diminishing anonymity), leading to increased mutual cooperation; and (2) a local adaptation mechanism wherein a player copies a strategy from a neighbor linked to the player through a network. These two effects explain how cooperators survive in a network PD game, even though players are required to use only the simplest strategies—either cooperation or defection (requiring only 1 bit of memory).[18] Therefore, hundreds of studies have investigated network reciprocity, primarily in the fields of theoretical biology and statistical physics.[19] The individuals of a population occupy the vertices of a graph. The edges denote who interacts with whom. Each individual interacts with all of their neighbors according to the standard payoff matrix, as given by Eq. (2.21). The payoff of each agent is totaled over all games with their neighbors. An individual's fitness is given by $1 - \omega - \omega F$, where F is the payoff for the individual and ω ($\omega \in [0, 1]$) is the intensity of selection. Here, we consider evolutionary dynamics according to Death–Birth updating (DB),[20] where, in each round, a random individual is chosen to die; then, their neighbors compete for the empty site in proportion to their fitness.

A calculation using pair approximation on regular graphs (with each vertex having k edges) leads to a deterministic differential equation that describes how the expected frequency of cooperation (defection) changes over time. This differential equation is actually a standard replicator equation with a modified payoff matrix.[21]

Based on the above discussion, now we are able to describe the five reciprocity mechanisms as the following transformed 2×2 matrices from the original game payoff matrix, given by Eq. (2.21)[10]:

[18] Nowak and May (1992).

[19] Due to space limitations, we can cite only five of these here: Hassell et al. (1994), Ebel and Bornholdt (2002), Santos and Pacheco (2005), Santos et al. (2006), Yamauchi et al. (2010).

[20] Ohtsuki et al. (2006).

[21] Ohtsuki and Nowak (2006).

$$\text{Direct Reciprocity}: \begin{array}{c} \\ C \\ D \end{array} \begin{array}{c} \begin{array}{cc} C & D \end{array} \\ \begin{pmatrix} \dfrac{R}{1-w} & S+\dfrac{wP}{1-w} \\ T+\dfrac{wP}{1-w} & \dfrac{P}{1-w} \end{pmatrix}; \end{array} \tag{2.28}$$

$$\text{Indirect Reciprocity}: \begin{array}{c} \\ C \\ D \end{array} \begin{array}{c} \begin{array}{cc} C & D \end{array} \\ \begin{pmatrix} R & (1-q)S+qP \\ (1-q)T+qP & P \end{pmatrix}; \end{array} \tag{2.29}$$

$$\text{Kin Selection}: \begin{array}{c} \\ C \\ D \end{array} \begin{array}{c} \begin{array}{cc} C & D \end{array} \\ \begin{pmatrix} R & \dfrac{S+rT}{1+r} \\ \dfrac{T+rS}{1+r} & P \end{pmatrix}; \end{array} \tag{2.30}$$

$$\text{Group Selection}: \begin{array}{c} \\ C \\ D \end{array} \begin{array}{c} \begin{array}{cc} C & D \end{array} \\ \begin{pmatrix} (n+m)R & nS+mR \\ nT+mP & (n+m)P \end{pmatrix}; \end{array} \tag{2.31}$$

$$\text{Network Reciprocity}: \begin{array}{c} \\ C \\ D \end{array} \begin{array}{c} \begin{array}{cc} C & D \end{array} \\ \begin{pmatrix} R & S+H \\ T-H & P \end{pmatrix}. \end{array} \tag{2.32}$$

Here, H in Eq. (2.45) is defined as follows:

$$H = \frac{(k+1)(R-P) - T + S}{(k+1)(k-2)}. \tag{2.33}$$

We assume that $k > 2$.

It is worth noting that, in Eq. (2.30), summations of two players, $R + R$, $P + P$, and $(S + rT)/(1 + r) + (T + rS)/(1 + r)$, are consistent with those of Eq. (2.21), $R + R$, $P + P$, and $S + T$.

From these equations, we can draw each set of conditions, wherein C and D become evolutionarily stable strategies (ESS) and the internal equilibrium as shown in Table 2.3. One important point to be addressed is that all these conditions and the internal equilibrium can be described only by D_g' and D_r' with the defined model parameters (i.e., we are no longer relying upon D_g and D_r).

Assuming the parameters $w = 0.1$, $q = 0.1$, $r = 0.1$, $m = 50$, $n = 500$, and $k = 12$, we obtain Figs. 2.5, 2.6, 2.7, 2.8, and 2.9, (Fig. 2.5; direct reciprocity; Fig. 2.6; indirect reciprocity; Fig. 2.7; kin selection; Fig. 2.8; group selection; Fig. 2.9; network reciprocity), wherein the D_g–D_r (upper) and D_g'–D_r' diagrams (lower) diagrams of equilibria for different $R - P$ are shown. Those results come from Eqs. (2.28, 2.29, 2.30, 2.31, 2.32, and 2.33), assuming that an initial cooperation fraction of 0.5. In these figures, white circles indicate the boundary point of four game classes: D-dominant, C-dominant, polymorphic, and bi-stable (hereafter "four-corners"). Each horizontal black broken-line in Figs. 2.5 and 2.6 means the

Table 2.3 Conditions for cooperation being ESS, defection being ESS, and interior equilibrium for direct reciprocity, indirect reciprocity, kin selection, group selection, and network reciprocity

	Cooperation is ESS	Defection is ESS	Internal equilibrium
Direct reciprocity	$\frac{w}{1-w} > D'_{\mathrm{g}}$	$D'_{\mathrm{r}} > 0$	$x^* = \frac{(1-w)D'_{\mathrm{r}}}{(1-w)\left(D'_{\mathrm{r}} - D'_{\mathrm{g}}\right) + w}$
Indirect reciprocity	$\frac{q}{1-q} > D'_{\mathrm{g}}$	$D'_r > 0$	$x^* = \frac{(1-q)D'_r}{(1-q)\left(D'_r - D'_{\mathrm{g}}\right) + q}$
Kin selection	$r(1 + D'_{\mathrm{r}}) > D'_{\mathrm{g}}$	$r\left(1 + D'_{\mathrm{g}}\right) < D'_{\mathrm{r}}$	$x^* = \frac{-r\left(D'_{\mathrm{g}} + 1\right) + D'_{\mathrm{r}}}{(1+r)\left(D'_{\mathrm{r}} - D'_{\mathrm{g}}\right)}$
Group selection	$\frac{m}{n} > D'_{\mathrm{g}}$	$\frac{m}{n} < D'_r$	$x^* = \frac{nD'_r - m}{n\left(D'_r - D'_{\mathrm{g}}\right)}$
Network reciprocity	$k^2 D'_{\mathrm{g}} - k\left(D'_{\mathrm{g}} + 1\right) + \left(D'_{\mathrm{r}} - D'_{\mathrm{g}}\right) < 0$	$k^2 D'_{\mathrm{r}} - k(D'_{\mathrm{r}} + 1) + \left(D'_{\mathrm{g}} - D'_{\mathrm{r}}\right) > 0$	$x^* = \frac{(k^2 - k - 1)D'_{\mathrm{r}} + D'_{\mathrm{g}} - k}{(k^2 - k - 2)\left(D'_{\mathrm{r}} - D'_{\mathrm{g}}\right)}$

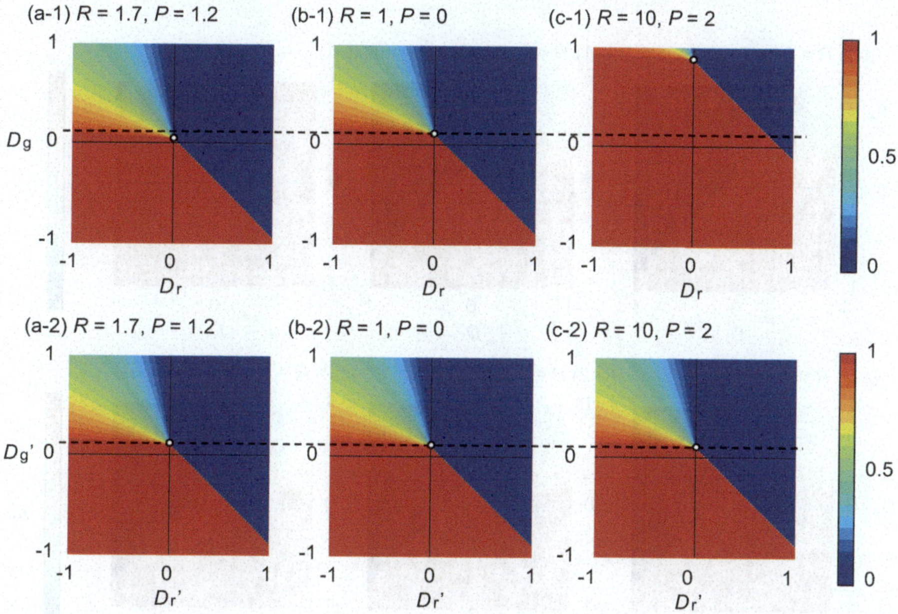

Fig. 2.5 Equilibrium-cooperation fraction-D_r–D_g (in the upper line) and D_r'–D_g' (in the lower line) diagrams of direct reciprocity for (**a**) $R = 1.7$, $P = 1.2$; (**b**) $R = 1$, $P = 0$; and (**c**) $R = 10$, $P = 2$, with the probability of another round given by $w = 0.1$

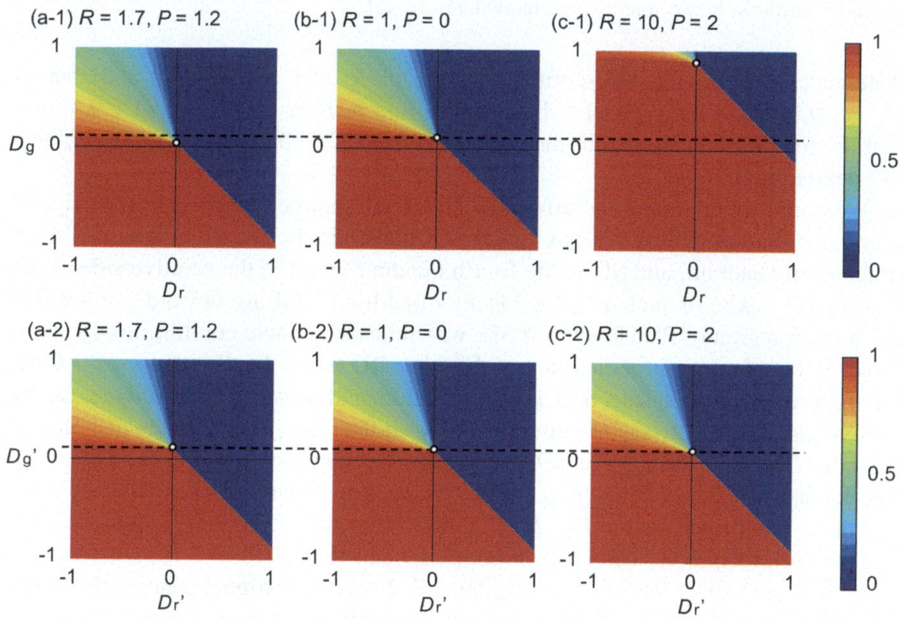

Fig. 2.6 Equilibrium-cooperation fraction D_r–D_g (upper line) and D_r'–D_g' (lower line) diagrams of indirect reciprocity for (**a**) $R = 1.7$, $P = 1.2$; (**b**) $R = 1$, $P = 0$; and (**c**) $R = 10$, $P = 2$, with the probability of knowing the reputation of another individual given by $q = 0.1$

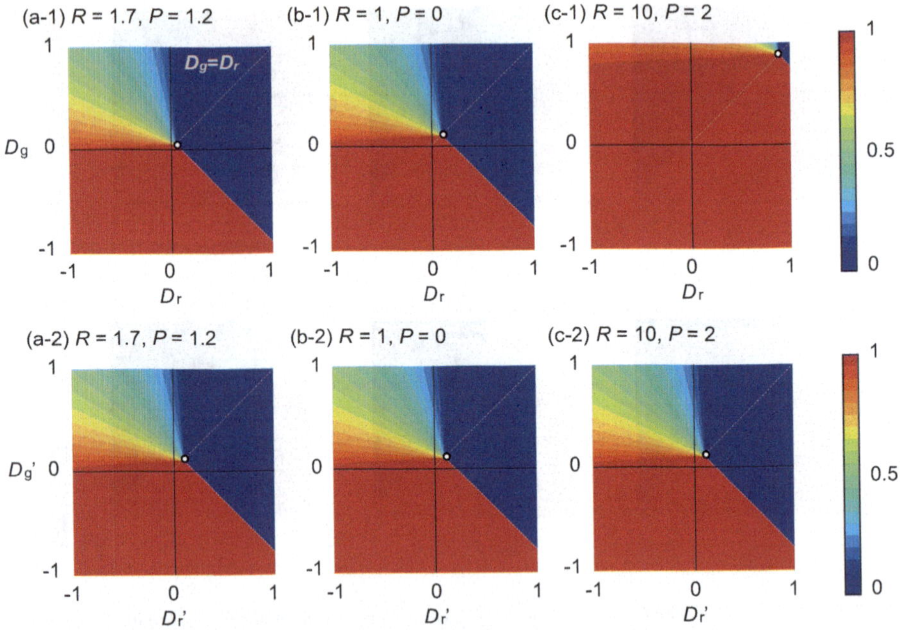

Fig. 2.7 Equilibrium cooperation fraction-D_r–D_g (in the upper line) and D_r'–D_g' (in the lower line) diagrams of kin selection for **(a)** $R = 1.7$, $P = 1.2$, **(b)** $R = 1$, $P = 0$, and **(c)** $R = 10$, $P = 2$ with the average relatedness between interacting individuals $r = 0.1$

difference between the four-corners in case of $R = 1$ and $P = 0$ and that of $D_g' = D_r' = 0$, whereas each white broken-line in Figs. 2.7, 2.8, and 2.9 shows that the four-corners shift from the point of $D_g' = D_r' = 0$ along the $D_g' = D_r'$ line (45 degree line).

We confirm that, under the effects of direct and indirect reciprocity, the original four game-classes (PD in the first quadrant, Chicken in the second quadrant, Trivial in the third quadrant, and SH in the fourth quadrant) shift to the positive side of the D_g-axis (D_g'-axis) (depicted by the black dotted line). Shifting upward implies that the weaker region of PD changes to SH, which has a bi-stable equilibrium. In short, direct and indirect reciprocities can weaken the GID ($D_g = T - R$ or $D_g' = (T - R)/(R - P)$). Furthermore, kin selection, group selection, and network reciprocity cause the four-original game-classes to simultaneously shift to the positive sides of both the D_g-axis (D_g'-axis) and D_r-axis (D_r'-axis). Shifting upper-right along the $D_g' = D_r'$ line means that a weaker region of PD changes to either Chicken, SH, or even Trivial, as confirmed by the D_g'–D_r' diagram. Thus, kin selection, group selection, and network reciprocity can weaken both GID and RAD.

In the D_g–D_r diagrams of Figs. 2.5, 2.6, 2.7, 2.8, and 2.9 (upper panels), the larger $R - P$ becomes, the more upward shifting can be observed. In the D_g'–D_r' diagrams of Figs. 2.5, 2.6, 2.7, 2.8, and 2.9, (lower panels), however, upward shifting occurs without regard for $R - P$. Surprisingly, the respective equilibria on the D_g'–D_r'

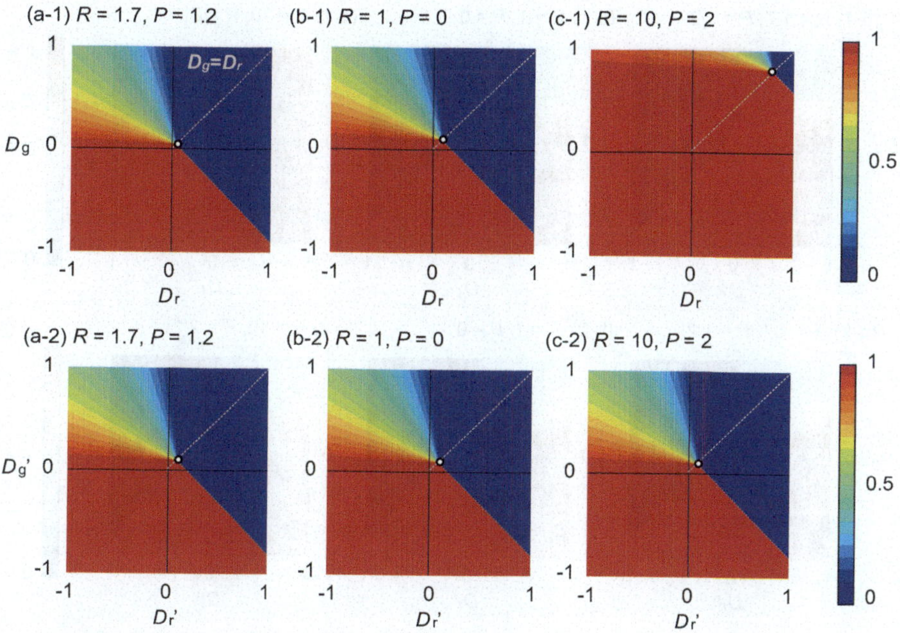

Fig. 2.8 Equilibrium-cooperation fraction D_r–D_g (upper line) and D_r'–D_g' (lower line) diagrams of group selection for (**a**) $R = 1.7$, $P = 1.2$; (**b**) $R = 1$, $P = 0$; and (**c**) $R = 10$, $P = 2$, with the number of groups $m = 50$ and maximum size of a group $n = 500$

diagrams are completely consistent with each other, despite having different $R - P$. Therefore, a set of parameters, D_g' and D_r', that considers D_g and D_r as well as $R - P$, is universally appropriate for evaluating dilemma strength in a population under any reciprocity mechanism.

Concerning kin selection (see Fig. 2.7), it is interesting to note that the cooperation fraction increases with both D_g' and D_r' (see the dotted box in Fig. 2.10a). Figure 2.10 shows equilibrium cooperation when assuming $r = 0.3$. Figure 2.10a shows the D_g'–D_r' diagram, and (b) shows the relationship between D_g' and the cooperation fraction when $D_r' = 0.46$. In a population with kin selection, the payoff of a focal agent is determined by a part of the opponent agent's payoff, as described in Eq. (2.30). This creates evolutionary dynamics attracted by an internal (polymorphic) equilibrium in which both C and D agents co-exist, rather than an equilibrium consisting of only D agents, because the mutual dependency on payoff enables an agent to offer C instead of D, even if increasing the risk of exploitation by an opponent. It is possible in this situation that the total of S and the contribution from the opponent (i.e., rT) will be much greater than $(1 + r)P$, and this possibility increases as r and T (D_g') become larger. This behavior explains the paradoxical situation whereby more cooperative equilibrium can be attained under a larger D_g'. This effect can be readily observed in real human societies.

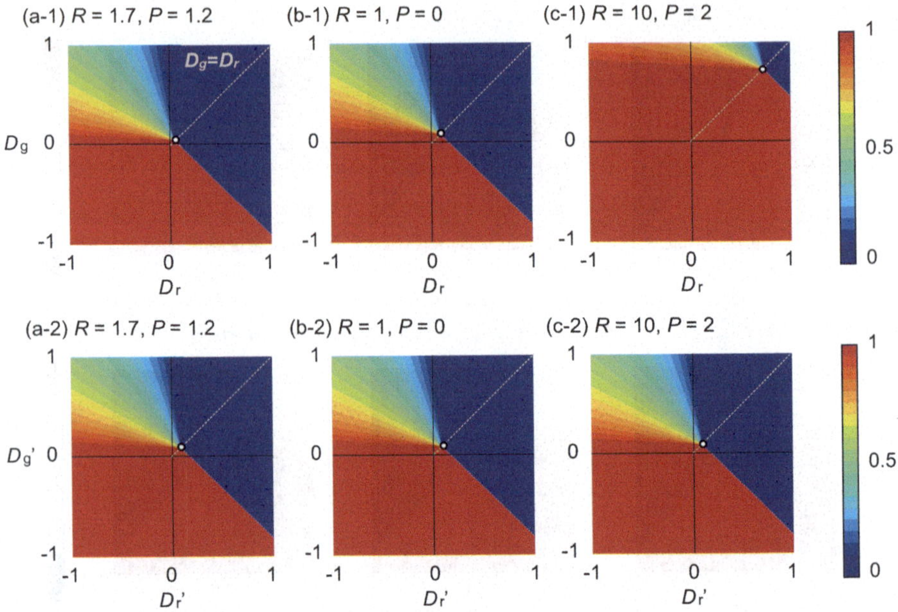

Fig. 2.9 Equilibrium cooperation fraction D_r–D_g (upper line) and D_r'–D_g' (lower line) diagrams of network reciprocity for (**a**) $R = 1.7$, $P = 1.2$; (**b**) $R = 1$, $P = 0$; and (**c**) $R = 10$, $P = 2$, with number of neighbors $k = 12$

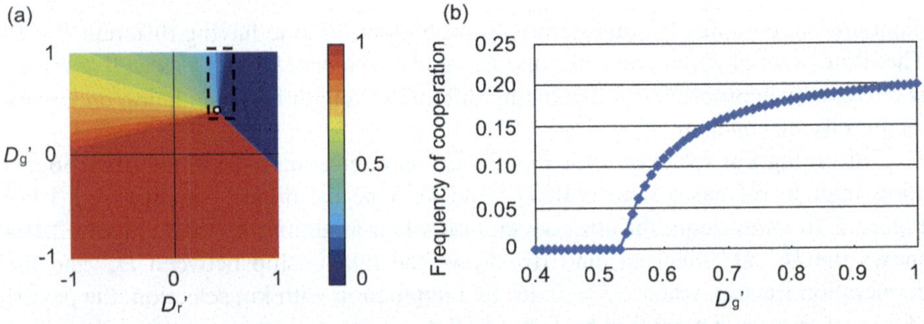

Fig. 2.10 Equilibrium cooperation fraction within the limits of (**a**) $-1 \leq D_g' \leq 1$ and $-1 \leq D_r' \leq 1$, and (**b**) $0.4 \leq D_g' \leq 1$ and $D_r' = 0.46$ (highlighted areas within dashed-line boxes), with ab average relatedness between interacting individuals given by $r = 0.3$

2.2 Multi-player Games

Though we have so-far assumed that there are two game players, a multi-player situation is more typical in a realistic context. It is therefore natural to extend our discussion to multi-player games. In fact, considering real applications of evolutionary game theory for applications to problems such as traffic flow and vaccination games (as we discuss later in this book), numerous agents may simultaneously play a social game from which decision-makers may extract information to efficiently regulate traffic flow (avoiding jams, for instance) or slow the spread of influenza. Thus, it is quite important to give a fundamental basis for quantifying multi-player games.

First, we outline the so-called public-goods Game (PGG), which has been used most often in the field as a template for multi-player games. This game is based on a social dilemma surrounding a public good that can only be sustained by a reasonable number of moral-minded cooperators through their donations. This means that players have an incentive not to donate (cooperate), but also to want to get their share of the cooperative fruits brought about by the donations of others.

Suppose G players participate in a single multi-player game, where a cooperator is requested to donate a cost c (in most cases, as in Fig. 2.11, assuming $c = 1$) to a public pool. We call G the game size. We assume that the entire population is infinite and well-mixed; however, at each gaming step, G players are randomly selected from the infinitely large mother population. Let the number of cooperators among G be n_c. After collecting the donations from all cooperators among the G players, the total pooled donation is multiplied by an amplifying factor, r. Thus, the public good is amplified. The fruits of this public good are distributed equally to all game participants, irrespective of whether they are a cooperator or a defector. In this sense, a defector can be called a free-rider.[22] Here, we can define the payoff-structure functions for both cooperators and defectors: $\pi_C(n_c)$ and $\pi_D(n_c)$, as shown in Fig. 2.11. The functions are defined by the number of cooperators in the group and can also be defined by the local cooperation fraction, $n_c/G \equiv P_C$. One important consideration is that the defectors' payoff is always larger than that of the cooperators at any particular cooperation fraction for a PGG. This schematic relation is redrawn more precisely in Fig. 2.12, where the cooperator and defector plots indicate the respective payoffs at the local cooperation fractions, $P_C = n_c/G$. The figure obviously suggests that, as long as $\pi_D(n_c - 1) > \pi_C(n_c)$ is satisfied, a cooperator has no incentive to keep cooperating at any cooperation fraction, and thus this fraction is always declining, regardless of its initial value. Consequently, Nash equilibrium is absorbed by an all-defectors state, i.e., $P_C = 0$. On the other hand, the maximum social payoff, or fair Pareto optimum, appears in the all-cooperators state, $P_C = 1$. This is why we can basically identify PGG as a multi-player prisoner's dilemma (N-PD) game.

One significant consideration is that the global-cooperation fraction, denoted by x, is different from the local cooperation fraction, P_C. Applying the concept of replicator dynamics (e.g., 2.7 or 2.22), we obtain the global dynamical equation as below:

---

[22]Speaking precisely, we should call them a first order free-rider, because in models considering punishment mechanisms, there are first order free-riders, meaning simple defectors, as well as second order free-riders, i.e., cooperators who are not punishing other defectors.

$$\dot{x} = x[f_C - (x \cdot f_C + (1-x) \cdot f_D)] = x(1-x) \cdot [f_C - f_D], \tag{2.34a}$$

$$f_C = \sum_{j=0}^{G-1} \binom{G-1}{j} x^j (1-x)^{G-1-j} \pi_C(j+1), \tag{2.34b}$$

$$f_D = \sum_{j=0}^{G-1} \binom{G-1}{j} x^j (1-x)^{G-1-j} \pi_D(j). \tag{2.34c}$$

As a simple case, let us assume that both payoff functions feature monotonic increase, as shown in Fig. 2.13. Solving Eq. (2.34), we can classify any game into

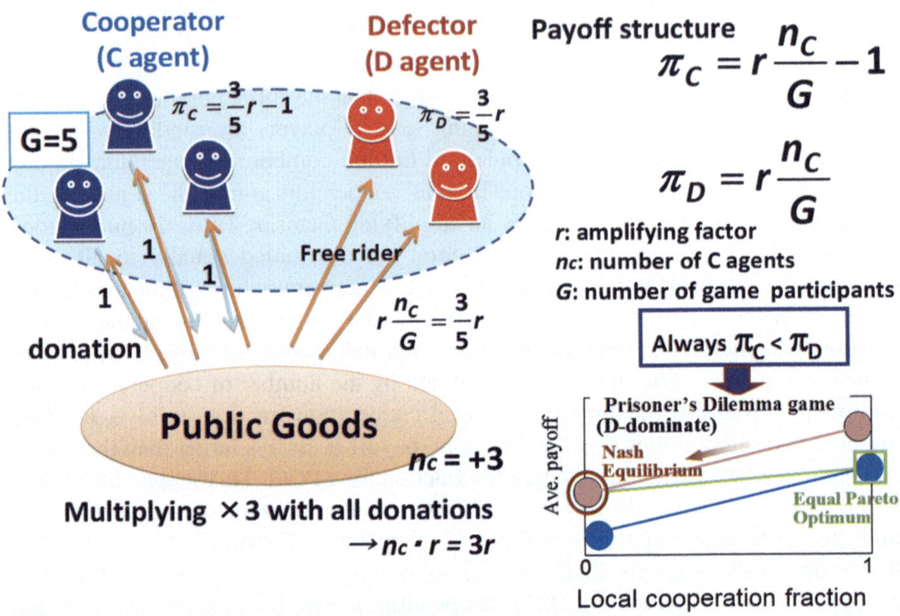

Fig. 2.11 Public-goods game (PGG); N-prisoner's dilemma game

Fig. 2.12 Payoff-structure function of the multi-player PD

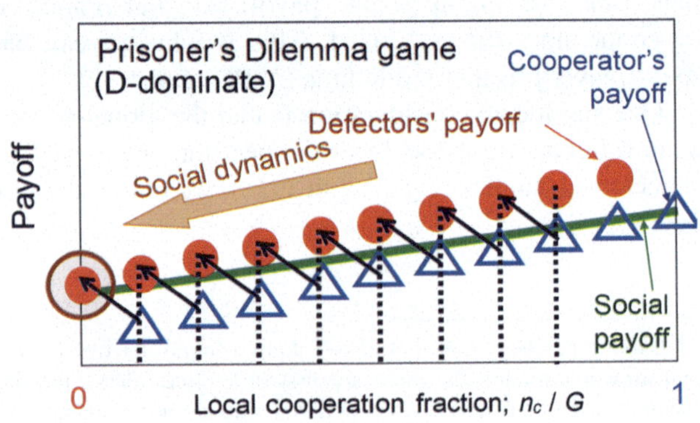

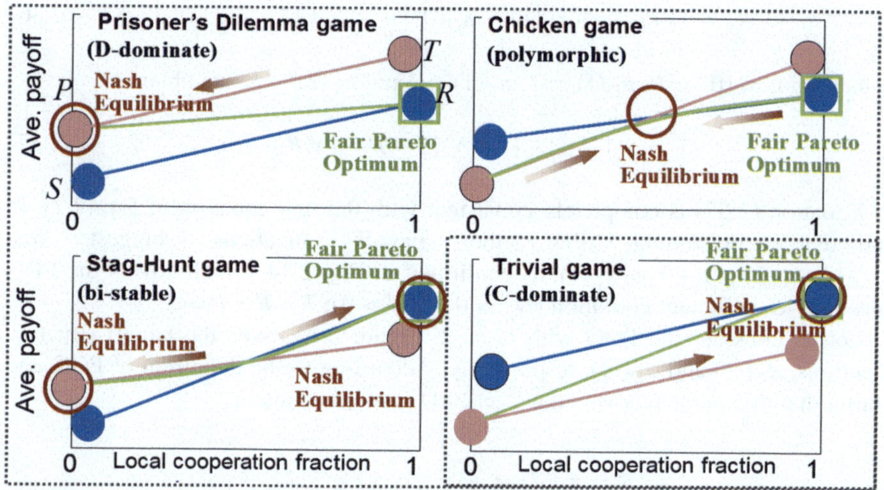

	C	D
Cooperation (C)	*R*, *R*	*S*, *T*
Defection (D)	*T*, *S*	*P*, *P*

Fig. 2.13 Four game classes and payoff-structure functions of multi-players games

one of four types, as in the case of 2×2 games: PD, Chicken, SH, and Trivial. Multi-player Chicken (N-Chicken) is featured when the cooperator's payoff function intersects the defector's at a certain local-cooperation fraction. Thus, a multi-player Chicken game has an internal equilibrium point, as in 2×2 Chicken. But note that the internal equilibrium is inconsistent with the crossing point of the payoff functions. Returning to Eq. (2.34), the internal equilibrium x_* must be obtained from

$$\sum_{j=0}^{G-1} \binom{G-1}{j} x_*^{\,j} (1 - x_*)^{G-1-j} [\pi_C(j+1) - \pi_D(j)] = 0. \tag{2.35}$$

In general cases, the number of x_* internal equilibria is not necessarily one but may be more, depending on the forms of both payoff functions. A multi-player Chicken game termed "the tragedy of the commons" has been accepted as one of the typical template models for describing a social dilemma caused by environmental problems. Multi-player Stag Hunt (N-SH) games also have a crossing point between the two payoff functions. Multi-player Trivial (N-Trivial) has no social dilemma, since cooperation dominates defection, meaning that the cooperator's payoff exceeds the defector's at any local-cooperation fraction.

Let us confirm that Eq. (2.34) recovers the 2×2 PD game when $G = 2$. In practice, Eq. (2.34) becomes

$$\dot{x} = x(1-x)[(1-x)(\pi_C(1) - \pi_D(0)) + x(\pi_C(2) - \pi_D(1))]. \tag{2.36}$$

Substituting $\pi_D(0) = P$, $\pi_D(1) = T$, $\pi_C(1) = S$and $\pi_C(2) = R$, we obtain

$$\dot{x} = x(1-x) \cdot [(1-x)(S-P) + x(R-T)]. \tag{2.37}$$

Equation (2.37) is completely consistent with the first equation of Eq. (2.7). In particular, presuming PD, the payoff functions suggest that $\pi_D(1) > \pi_C(2) > \pi_D(0) > \pi_C(1)$, as confirmed in Fig. 2.14, which covers the PD's necessary & sufficient condition ($D_g{}' > 0$ & $D_r{}' > 0$); $T > R > P > S$.

Let us consider the PGG with $G = 2$. Figure 2.14 gives the payoff-structure functions, $\pi_C(n_c)$ and $\pi_D(n_c)$, respectively. According to the definition of PGG and noting that this game recovers the 2 × 2 PD game, we obtain

$$R = \pi_C(2) = \frac{2c}{2}r - c = c(r-1),$$

$$T = \pi_D(1) = \frac{c}{2}r,$$

$$S = \pi_C(1) = \frac{c}{2}r - c,$$

and

$$P = \pi_D(0) = 0,$$

where $1 < r < 2$ to make sure $T > R > P > S$. One may conceive of this particular game as being a donor and recipient (D & R) game. As discussed before, a D & R game features $R = b' - c'$, $T = b'$, $S = -c'$, and $P = 0$, where c' and b' respectively indicate the "cost for cooperation" and "benefit of cooperation". More to the point, we should

Fig. 2.14 Payoff-structure functions $\pi_C(n_c)$ and $\pi_D(n_c)$ in the case of $G = 2$, public-goods game

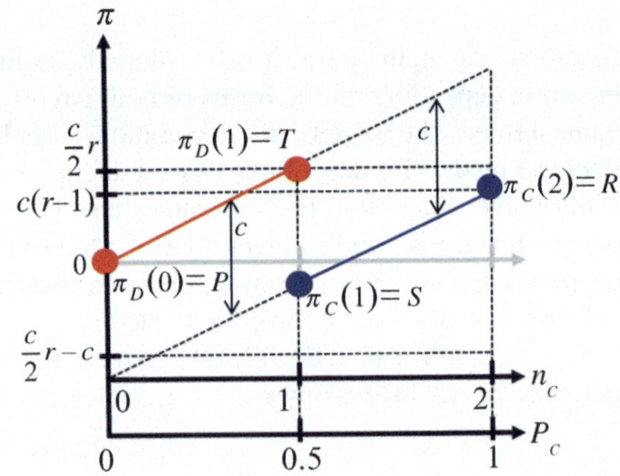

note that $D_g = D_r = c'$ $(D'_g = D'_r = \dfrac{c'}{b' - c'})$, whereas in this PGG, the dilemma strength is $D_g = D_r = c - \frac{c}{2}r$ $(D'_g = D'_r = \dfrac{1 - r/2}{r - 1})$, which implies parameter r denotes the dilemma weakness. Hence, a smaller r causes the dilemma to increase. Consequently, those two games are substantially similar under application of the transformations $c' = c - \frac{c}{2}r$ and $b' = \frac{c}{2}r$. In sum, a PGG, defined as multi-player game, is substantially equivalent to a D & R game, given as a 2 × 2 game.

If G becomes quite large, the local cooperation fraction will be approximately consistent with the global cooperation fraction. Thus, the mean-field approximation (MFA), $x \cong P_C = n_c/G$, transforms Eq. (2.34) into.

$$\begin{aligned}\dot{P}_C &= P_C[\pi_C(P_C) - (P_C \cdot \pi_C(P_C) + (1 - P_C) \cdot \pi_D(P_C))] \\ &= P_C(1 - P_C) \cdot [\pi_C(P_C) - \pi_D(P_C)]. \end{aligned} \tag{2.38}$$

In this case, the crossing point of the payoff functions yields the internal equilibrium for multi-player Chicken and SH.

2.3 Social Dilemma Alleviated by Integrated Reciprocity Mechanisms

In this section, we will see a state-of-the-art result in the field of evolutionary game theory (EGT). Aside from the analytical approach we have taken so far, a simulation approach is introduced here. For last several decades, the portentous growth of computation resources, through hardware as well as software, has pushed forward what is called multi-agent simulation (MAS), resulting in numerous studies based on huge sets of systematic simulations, which could never have been performed before. MAS is quite flexible and useful because it is applicable to any dynamical system for which the set of governing equations is unknown. Besides MAS, a conventional numerical approach always requires a set of governing equations describing the dynamics of a concerned system. In fact, computational fluid dynamics (CFD), one of the most frequently used engineering tools, is governed by the Navier–Stokes equations (N-S equations), which describe the time-evolutionary balance of momentum, energy, and other scalars at an infinitely small volume. On the other hand, MAS only requires a set of local rules regulating the dynamics of each agent, rather than global rules given by a set of governing equations. In this sense, MAS seems quite appropriate for introduction to an EGT model considering more realistic and more complex situations that are not covered by the analytical approach, such as the application of replicator dynamics.

As below, we are concerned with Nowak's five reciprocity mechanisms, especially network and indirect reciprocity. Our major motivation is seeing what would happen if network reciprocity were to be integrated with indirect reciprocity into a single EGT model—i.e., whether it would result in an enhanced cooperation less or greater than what would be expected by simply summing the individual contributions of the two mechanisms.

2.3.1 *Motivation and Background*

The emergence of altruistic behavior and how it can prove sustainable among many animal species has attracted the interest of many researchers including biologists and statistical physicists, because paying costs for cooperation to help others is at odds with natural selection. EGT, to which a huge amount of computational and experimental effort has been directed, has provided an array of possible answers to this challenging question. One set of explanations involves assortment caused by a spatially structured population, which is called network reciprocity. Although EGT may assume either 2×2 games or multi-player–2-strategy games without loss of generality, PD games, tend to be accepted as a good metaphor for social-dilemma situations.

Network reciprocity goes back to a pioneering work by Nowak & May[23] on the spatial prisoner's dilemma (SPD) game. Since then, network reciprocity has been the subject of a great deal of study both theoretically and numerically,[24] as well as experimentally.[25] Although the central assumption of the model, namely, "playing with the neighbors and copying successful strategies from them," is simple, the associated models show that even agents that are unsophisticated in terms of information processing can develop cooperative social systems. But one question that could be raised is whether more intelligent agents, such as humans, really rely on such a simple mechanism. It seems plausible that one would react differently to each neighbor depending on their particular cooperative tendency, rather than just uniformly either cooperating (C) or defecting (D) with all neighbors, as most previous SPD models have assumed. An intelligent agent can use information derived from the observation on what happened around them the previous time to draw an appropriate assessment of whether their neighbors act cooperatively. This information allows them to react in a more sophisticated and complex manner, and may eventually lead to a higher payoff. Thus, a cleverer means of information use could evolve.

The term "information" contains several layers. Anything helping an agent to distinguish someone from others can be viewed as "information". In this sense, what is called tag recognition provides information.[26] However, tags are constant or do

[23]Nowak and May (1992).

[24]Because of spatial limitations, we can only cite a few of these here: Szabo and Fath (2007), Perc et al. (2013), Perc (2010), Wang et al. (2015b).

[25]Although there have been many such studies, a representative work is given by Fehl et al. (2011).

[26]I would suggest two representative works: Riolo et al. (2001), Roberts and Sherratt (2002).

not change frequently over an agent's lifetime. Although tag systems can help an agent to evaluate their similarity to opponents, they are only of temporary use, since tags can be expected to lead to continuous adaptation.[27] By contrast, a mechanism providing time-variable information that can indicate the cooperative tendency of a co-player could foster emerging cooperation. This would be a mechanism of indirect reciprocity. Indirect reciprocity can be intuitively understood from the phrase "I will help you if you have helped someone".[28] There have been many brilliant studies on indirect reciprocity.[29] These models are based on two ingredients: the first is *reputation* which indicates whether one's opponent is good (hereafter, G) or bad (hereafter, B) – or, alternatively, another, finer assessment of players. The second concept is a *norm*, which defines what is good or bad based on what happened in the past. Thus, the norm is a rule to assess one's reputation. Apart from a few exceptions,[30] a single norm is shared with the whole population in these models; otherwise, emerging cooperation becomes very fragile.[31] A norm determines whether a certain agent is good or bad based on their actions towards their opponent and this opponent's reputation. Additionally, in some cases,[32] the focal agent's reputation—either G or B—is also taken into account. The number of norm variations is $(2^2)^2 = 16$ for the former case or $((2^2)^2)^2 = 256$ for the latter case.

What would be observed if two different mechanisms fostering the evolution of cooperation are combined into a single model? Since there are two mechanisms, does this enhance cooperation even more? Or, because of a certain negative-interference effect, does it rather deteriorate what each single mechanism could bring? Several studies have assumed network reciprocity and combined it with another system, such as direct reciprocity. For example, Miyaji et al.[33] reported that more cooperation can be established by additionally presuming direct reciprocity in an SPD game, where an agent in a spatially structured system is allowed to have memory and a strategy to distinguish each of their neighbors when they cooperate or defect. On the other hand, combining a tag-recognition model with a structured population has shown that how high and low assortment of a population affect the tag-recognition mechanism in very different ways, sometimes even effectively decreasing cooperation.[27] Nowak & May[23] have clarified that combining a tag model with network reciprocity does not lead to positive but rather to negative interference in terms of cooperation.

However, the question of what would happen if an agent who is playing SPD games is given information based on an indirect-reciprocity mechanism is still open.

---

[27]Garcia et al. (2014).

[28]Nowak and Sigmund (2005).

[29]For readers' convenience, the most representative works are given below: Sugden (1986), Wedeking and Milinski (2000), Panchanathan and Boyd (2004).

[30]Chalub et al. (2006).

[31]Tanimoto and Sagara (2015).

[32]We suggest the most representative works: Ohtsuki and Iwasa (2004, 2006).

[33]Miyaji et al. (2013).

Thus, the present study seeks to answer whether or not adding indirect reciprocity to network reciprocity can foster cooperation. Although one may think that this should certainly increase cooperation compared to conventional SPD games, more information could also devastate network reciprocity. The current study intends to address this particular question by building a minimal model dovetailing network reciprocity with indirect reciprocity.

2.3.2 Model Setup

First, we assume a simple network-reciprocity model as the default case, where synchronous Imitation Max (hereafter IM; otherwise known as "best-takes-over"[34]) is assumed as an agent's strategy-updating rule. As a population structure, we assume a 2D lattice with a degree of 8 ($k = 8$), i.e., a Moore neighborhood. This is a very popular deterministic-update rule, whereby an agent imitates the neighboring strategy that lead to the highest payoff among both his neighbors and himself. Since we aim for a minimal model, no stochastic elements that might arise through asynchronous updating, stochastic-updating rules, or heterogeneous networks, are taken into account. This is because the introduction of such stochastic elements inevitably makes it more difficult to draw a conclusion to the above question.

In the default setting, we focus upon two cases. One assumes a discrete-strategy system in which an agent has a binary strategy of either C or D. A second case assumes a mixed-strategy system in which an agent uses a real number between 0 (D) and 1 (C) as his strategy, but the action is still restricted to either C or D; therefore, their actions are probabilistically determined by strategy.

In our model, a focal agent, *i*, is able to obtain their neighbor *j*'s information, Obs_j, which is a binary *evaluation* of *j* as either good (G) or (B). On the other hand, each agent has a *strategy*, defined by *Str*, which evolves in time according to synchronous IM. The action of agent *i* towards neighbor *j* is defined as.

$$\text{If } Obs_j \geq Str_i \text{ Agent } i \text{ cooperates with Agent } j; \text{ otherwise defects.} \tag{2.39}$$

The evaluation of each agent; *Obs* is defined by *action assessment*. An important difference from previous indirect-reciprocity models[31,35,36] wherein a social norm determined an agent's reputation, is that the action assessment only refers to the actions of the opponent player in the previous time step, but not to the assessment of the focal and opponent players. We assume three types of action-assessment rules:

[34]Hauert (2002).

[35]Leimar and Hammerstein (2001).

[36]Pachanathan and Boyd (2003).

(a) **Action scoring (hereafter; AS):** if an agent cooperates, their action is evaluated as good (G). If they defect, they are labeled as bad (B).
(b) **Action judging (hereafter; AJ):** If an agent only cooperates with a cooperator or defects to a defector, their action is evaluated as G. Otherwise, it is evaluated as B.
(c) **Discourage exploitation (hereafter; DE):** An agent's action is assumed to be G as long as they are not exploiting their neighbor, i.e., as long as they do not defect against a cooperator .

Table 2.4 summarizes how each of the three action assessments are evaluated, leading to either G or B in a 2 × 2 game.

Note that AS only sees how a player behaves, i.e., whether they cooperate or defect, irrespective of their opponent's action. At this point, AS shares the same basic concept as what is called *Scoring*, one of the social norms,[37] which assesses those who give (refuse) help to good (bad) opponents, as well as tit-for-tat (TFT), a direct-reciprocity strategy[38] by which one cooperates with a cooperator and defects from a defector.

AJ encourages equal actions, obtaining R or P. The concept of AJ seems analogous to *Stern-judging*, which regards a defection for a bad opponent as good

Table 2.4 How each of the three action assessments yields either good (G) or bad (B) for mutual cooperation (R), mutual defection (P), Sucker (S) and Temptation (T) in a 2 × 2 game

(i) Action Scoring; AS

		Opponent	
		Cooperation	Defection
Focal player	Cooperation	G	G
	Defection	B	B

(ii) Action judging; AJ

		Opponent	
		Cooperation	Defection
Focal player	Cooperation	G	B
	Defection	B	G

(iii) Discourage exploitation; DE

		Opponent	
		Cooperation	Defection
Focal player	Cooperation	G	G
	Defection	B	G

[37] Nowak and Sigmund (1998), Brandt and Sigmund (2005).

[38] Axelrod (1984).

[39] Pacheco et al. (2006).

[40] Kraines and Kraines (1988).

(unlike Scoring),[39] and is qualitatively similar to Win-stay & Lose-shift,[40] which is another direct-reciprocity strategy by which a player cooperates after mutual defection as well as mutual cooperation and defects after being exploited as well as exploiting.

DE is partially analogous to Simple-standing, which assigns a good reputation as long as a player not exploit a good-labelled opponent.[35,36]

With respect to the information actually given to each agent (i.e., the observed data for each agent), *Obs*, we proceed on the following way:

> If the number of Agent *i*'s "good" actions towards their neighbors in the previous time step is less (more) than 4 ($= k/2$), Obs_i is 0 (1). If the number of good actions is 4, Obs_i is either 0 or 1 with equal probability. *Str* is then either 0, 1 or 2, respectively, meaning AllC, Discriminator (hereafter Disc) or AllD actions.

This implies that *Obs* does not directly indicate a precise evaluation of each neighbor, but only approximates information as G or B (we later mention the case of complete information).

The number of agent's strategy; *Str* is the number of *Obs* plus 1, e.g., $2 + 1 = 3$, in order to cover the full strategy space from all defection (AllD) through discriminator (Disc) to all cooperation (AllC). Thus, $Str \in \{0, 1, 2\}$.

Table 2.5 summarizes the relationship of agent's actions and *Str* versus *Obs*.

Concerning the initial setting in the following simulation, each agent is randomly assigned $Str = 0$, 1, or 2 with equal probability ($= 1/3$). The initial values of *Obs* in an agent's action profile for a first game are set to good (=1) for an AllC agent ($Str = 0$) and bad (=0) for an AllD agent ($Str = 2$); for a discriminator, *Obs* is either good or bad with equal probability.

Our simulations are based on SPD games. In a PD game, a player receives a reward (R) for mutual cooperation and a punishment (P) for mutual defection. If one player chooses C and the other chooses D, the latter obtains a temptation payoff (T), while the former obtained the sucker's payoff (S). We assume a spatial game with $R = 1$ and $P = 0$, parametrized as

Table 2.5 Action table of strategy (*Str*) in relation to the observed data (*Obs*) in the low-resolution case

			Str		
Number of observed "Good"	*Obs*		0	1	2
			AllC	Discriminator (Disc)	AllD
0	0 (=Bad)	Bad	C	D	D
1					
2					
3					
4	50%; 0/50%; 1	Good	C	If Good then C; else D	D
5	1 (=Good)		C	C	D
6					
7					
8					

$$\begin{pmatrix} R & S \\ T & P \end{pmatrix} = \begin{pmatrix} 1 & -D_r \\ 1 + D_g & 0 \end{pmatrix}, \tag{2.40}$$

where $D_g = T - R$ and $D_r = P - S$ imply a chicken-type dilemma and stag-hunt dilemma, respectively.[3,8] We limit the PD game class by assuming $0 \le D_g \le 1$ and $0 \le D_r \le 1$.

We set a population of size $N = 10^4$. In the following discussion, we are mainly concerned with the global fraction of cooperation, which is measured by the fraction of cooperation in the entire population at a certain time step. Although exploring the distributions of *Str* and *Obs* would yield a deeper understanding of the dynamical system, for example in terms of phase transitions in spatial games,[41] we focus on the global cooperation fraction because it provides a good index for exploring how cooperative a society may become under given assumptions. Global cooperation is defined as the real number of actions whether each of agents cooperates or defects.

2.3.3 *Results and Discussion*

2.3.3.1 Synoptic Result

Figure 2.15 shows the average fraction of cooperation for both default cases, assuming discrete and mixed strategy systems, respectively. Figure 2.16 indicates results presuming each of the three action assessments. In the AJ case, the average cooperation fraction as well as its standard deviation from 100 realizations are shown in Fig. 2.16a, b. Since AS and DE display only a small deviation amid 100 realizations, their standard deviations are not shown. Obviously, we can confirm that the mixed-strategy setting shows better performance than the discrete strategy does. Although we have made a precise study on this point,[42] the cause is summarized as below. We should consider the region of enhanced cooperation in the mixed strategy divided into two regions: boundary games to Chicken (BCH) and boundary games to Stag Hunt (BSH). In BCH games with mixed strategies, mid-cooperative clusters consisting of mid-cooperative strategies, which can establish reasonable cooperation, are formed; on the other hand, in BSH games with mixed strategy, a "come-from-behind" victory event, where a more cooperative agent defeats a defective agent, occurs probabilistically, causing a higher cooperation level than when occur adopting discrete strategy. In the following discussion, we primarily compare the present model with the mixed strategy, because our setting allows an agent to offer C or D to different neighbors, which is exactly same as the mixed strategy allowing.

---

[41]Traulsen and Claussen (2004).

[42]Zhong et al. (2012), Kokubo et al. (2015).

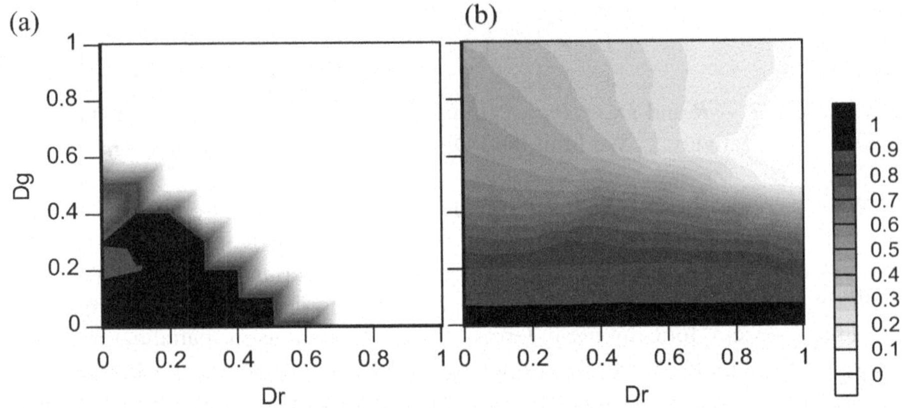

Fig. 2.15 Social average of cooperation fraction within the limits $0 \leq D_r \leq 1$ and $0 \leq D_g \leq 1$. All contours are drawn from an average of 100 realizations. Panel (**a**) default setting; conventional SPD assuming a discrete-strategy system, and (**b**) another default setting; conventional SPD assuming a mixed-strategy system

First, one notices that giving information concerning neighbors does not always enhance cooperation compared with the default case. In a nutshell, adding an indirect-reciprocity mechanism to network reciprocity, in some cases, decreases cooperation.

The AJ setting (Fig. 2.16a) shows the highest levels of cooperation, and outperforms the default cases (Fig. 2.15a, b). However, the AS (Fig. 2.16c) and DE settings (Fig. 2.16d) show only meager levels of cooperation, particularly in the chicken-type dilemma (area with larger D_g) if we compare them with the SPD with a mixed strategy setting (Fig. 2.15b). This will be discussed later.

2.3.3.2 Time Evolution

Figure 2.17 shows all cooperation-fraction time series across 100 realizations presuming AJ with $D_g = 0.6$ and $D_r = 0.9$, as highlighted by the plus mark in Fig. 2.16a, b. It represents an area of relatively large deviation over 100 realizations, as shown in Fig. 2.16b. 99 out of 100 trials show almost the same cooperation level in equilibrium. In a single case, the system is absorbed into the all-defectors-state. This results from the stochasticity arising from the inital condition.

Figure 2.18 shows time evolutions at three dilemma strengths: $D_g = D_r = 0.3$, 0.7, and 0.9, highlighted by the circle, square, and downward-triangle symbols in Fig. 2.16a, b. Each represents a typical episode out of 100 realizations. Each upper panel indicates the time evolutions of the cooperation fraction (black), average *Str* (blue), and average *Obs* (red). Likewise, each lower panel shows fractions of six classes composed of the agent's strategy (one of {AllC, Disc, AllD}) and their own *Obs* classified as either good or bad.

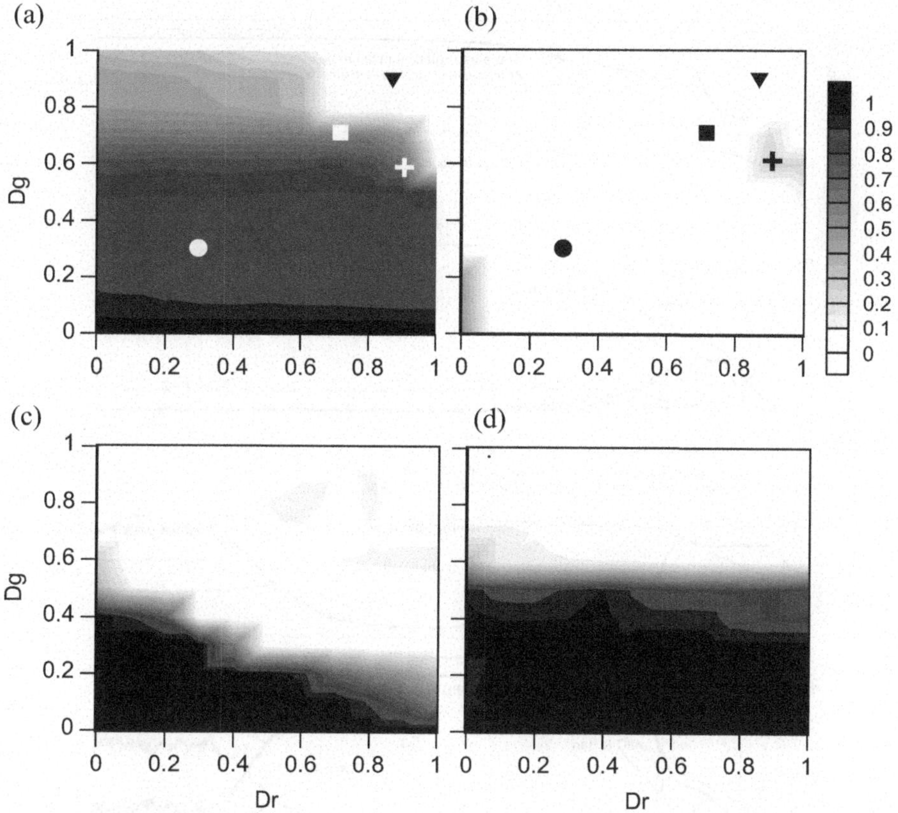

Fig. 2.16 All contours are formatted in the same manner as Fig. 2.1. Results when low resolution is given: (**a**) average cooperation fraction assuming an AJ setting; (**b**) standard deviation among 100 realizations of cooperation fraction assuming an AJ setting, and (**c**, **d**) average cooperation fractions assuming AS and DE settings, respectively

Fig. 2.17 All 100 time evolutions of 100 realizations, presuming a low resolution and AJ settings at $D_g = 0.6$ and $D_r = 0.9$, highlighted by the plus mark in Fig. 2.16a, b

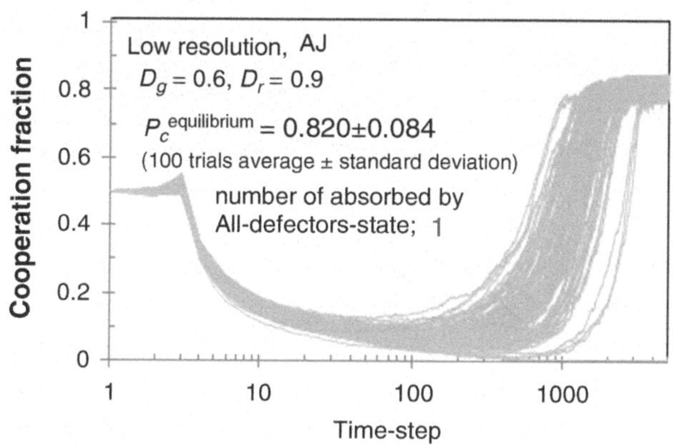

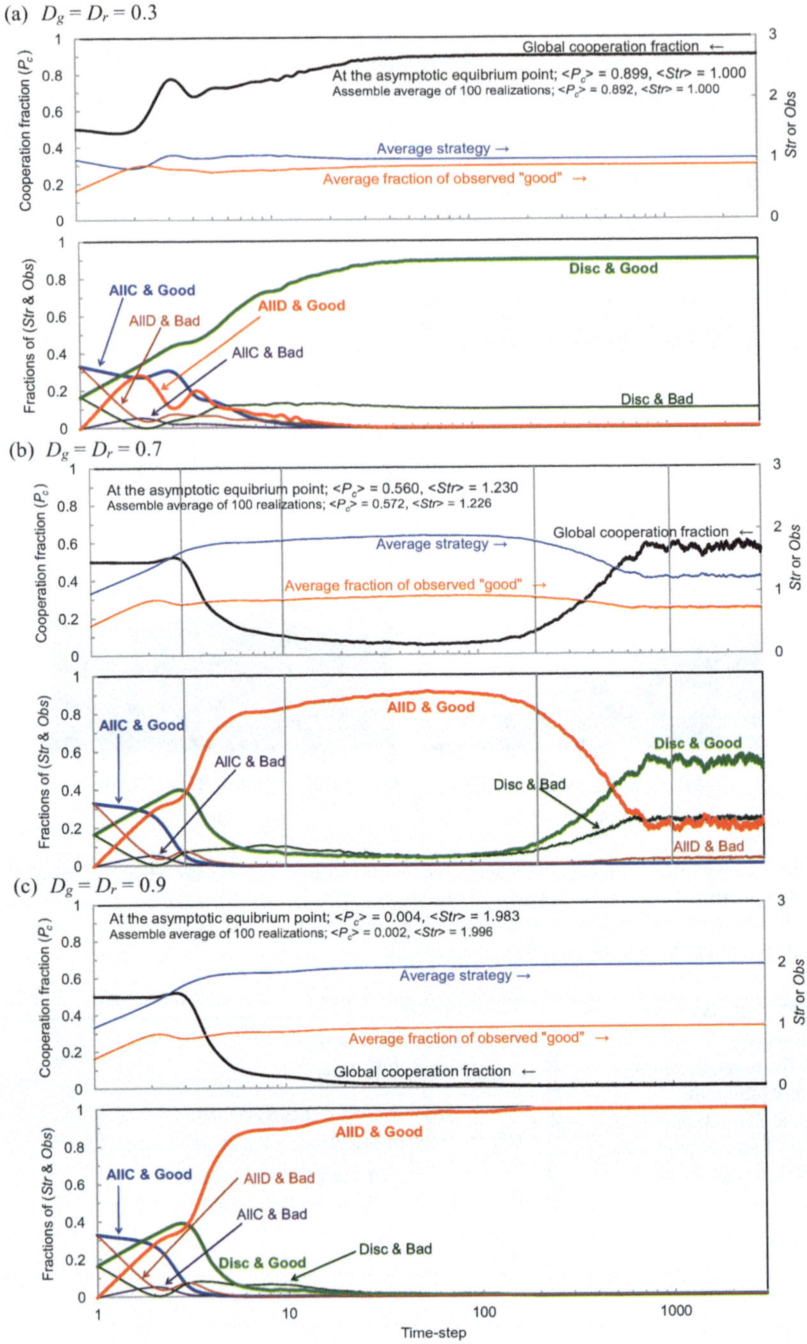

Fig. 2.18 Representing time evolutions of the three dilemma strengths presuming AJ shown by three symbols in Fig. 2.16a. Each upper panel shows a cooperation fraction (black), average *Str* (blue), and average *Obs* (red). Each lower panel shows the fractions of six classes combined by {AllC, Disc, AllD} and {Bad, God}

Obviously, in the case of a stronger dilemma game; $D_g = D_r = 0.9$ (panel (c)), AllD & Good agents become dominant over the other five classes at an early stage, giving discriminators no chance to survive cooperation. In fact, Disc & Good agents manage to survive for a substantial number of time-steps, before finally dying out.

On the other hand, in the case of a weaker dilemma game, $D_g = D_r = 0.3$ (panel (a)), Disc & Good agents successfully dominate others.

Those two cases display simple dynamics that allow a single dominant situation (AllD & Good or Disc & Good) to be attained at equilibrium, despite starting from an initially random situation. This implies that the dynamical system stipulated by those settings allows one single dominant state; either cooperative or defective. When the dilemma becomes stronger, the probabilty of ending in a cooperative state decreases.

However, for a dilemma of middle or reasonably large strength, which is the case for $D_g = D_r = 0.7$ (panel (b)), the dynamics do not allow a single dominant state at equilibrium, but a Disc & Good or AllD & Bad coexisting state that realizes a reasonably high cooperation level in a whole society (observed in the upper panel of Fig. 2.16a). This is discussed in detail in the next section.

2.3.3.3 Detailed Time Evolution Presuming AJ Settings with a Reasonably Large Dilemma

Figure 2.19 shows snapshots at 3, 10, 200, and 1000 time steps in the episode shown in Fig. 2.18b. The four gray lines in Fig. 2.18b indicate 3, 10, 200, and 1000 time steps.

Based on Figs. 2.18b and 2.19, the mechanism of maintaining a high cooperative level is relevant to the question of how a discriminator's cluster (hereafter, a Disc-cluster) survives in a sea of defectors. It seems analogous to the investigation of what mechanism can establish network reciprocity by Tanimoto et al.[43] Their question may shortly be phrased as how a cluster of cooperators (hereafter a C-cluster) survives several initial time steps starting from a completely random and chaotic situation (which is called the END period by Ref [44]), and to what extent the surviving C-clusters in the END period can extend into the neighboring defective field (which is called an EXP period by [44]). Assuming an IM with a lattice network as in this study, the END period may imply a relaxing-time starting from a unique stochastic-randomness imposed by an initial spatial distribution. If an episode is absorbed by the all-defectors-state, its time-evolution stops there. If cooperation successfully survives by forming appropriate C-clusters, the episode shits to the EXP period, where those C-clusters extend to a neighboring area. Ideally, only a single perfect C-cluster survives in END, namely a 3×3 cooperators block; this configuration is most efficient for obtaining a higher cooperation fraction in the EXP, which is basically able to attain an all-cooperators-state. If several C-clusters survive

[43]Wang et al. (2013).

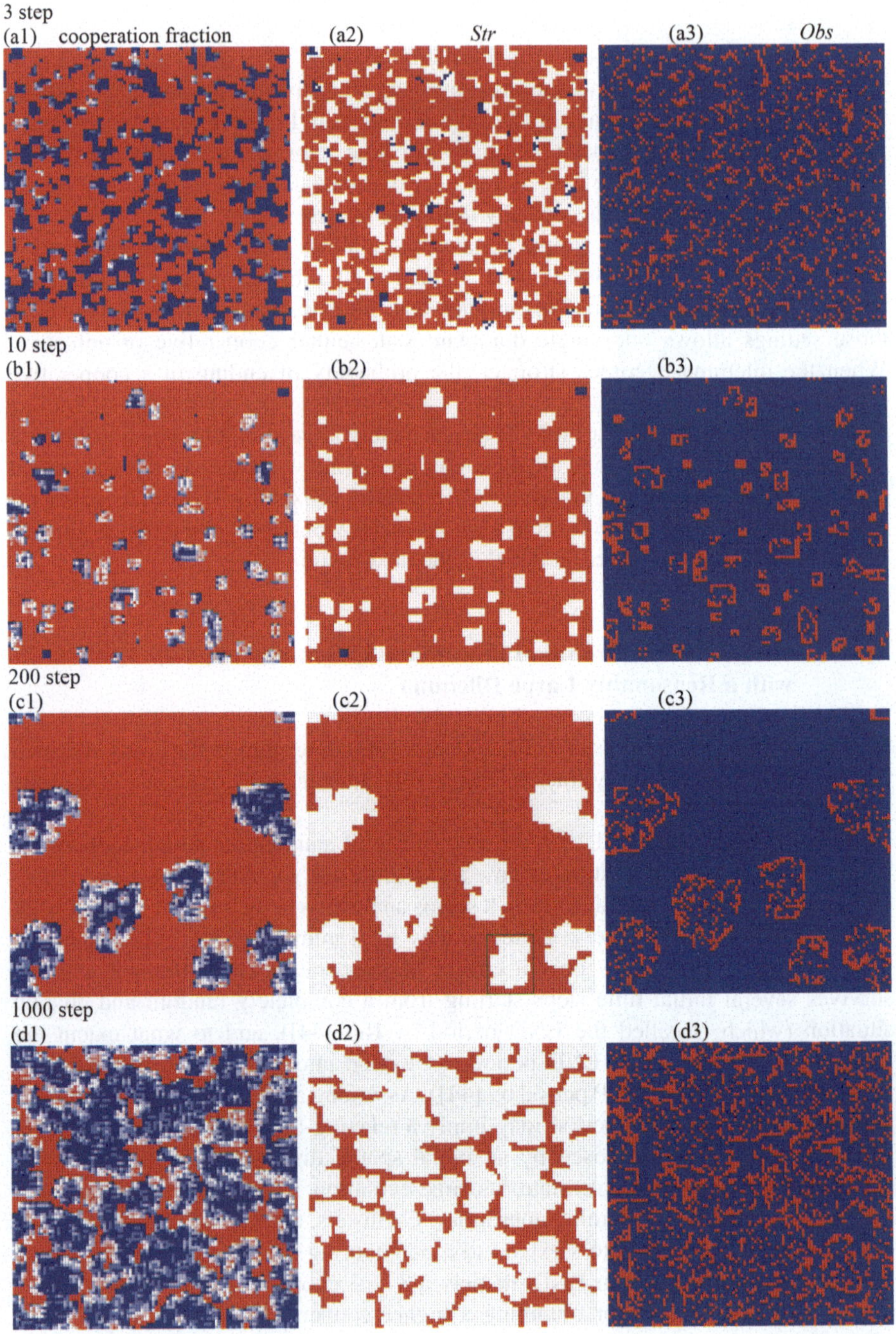

Fig. 2.19 Snapshots of the episode shown in Fig. 2.18b. Panels (**a***), (**b***), (**c***) and (**d***) show results of 3, 20, 200, and 1000 time steps, highlighted by gray vertical lines in Fig. 2.18b. Panels (*1), (*2), and (*3) indicate the cooperation fraction, *Str*, and *Obs* of each agent. Throughout all panels except (*3), blue, red, and white indicate cooperative, defective and neutral. For *Obs*, good (bad) is blue (red)

in END shift to EXP, several defectors might be able to survive at equilibrium in gaps between extended C-clusters by exploiting several cooperators.

As we will confirm in the following part, our model goes along with almost the exact scenario explained in the previous paragraph, only with discriminators and defectors instead of cooperators and defectors playing the main role in the dynamics.

Figure 2.18b indicates that Disc can increase after the two initial time steps, but with AllD increasing more rapidly. Thus, after 3 time steps, AllD & Good become the majority instead of Disc & Good. Disc & Good goes into decline after 3 time steps. Snapshots at 10 and 200 time steps indicate that most of the Disc-clusters surviving for the 3 initial time steps go to die in the sea of AllD, but a small number of those (i.e., 5 Disc-clusters as observed in Fig. 2.19c*) can survive and thereafter expand to the neighboring defective area. This is, as we depicted above, qualitatively the same for the END and EXP processes. In short, the early period of time evolution selects several robust Disc-clusters that are able to endure defectors invasion. Subsequently, the surviving Disc-clusters can expand to the neighboring defector area, leading to a reasonable co-existence of cooperation and defection at equilibrium. As explained above, in fact, at the 1000th time step, AllD agents survive in gaps between plural Disc-clusters, because they can exploit Disc agents, and maintain their defective strategy.

The next obvious question is why those five Disc clusters are ultimately able to survive and expand, unlike other Disc clusters.

Figure 2.20 shows detailed snapshots at 199 and 200 time step, giving the enlarged image of the green-highlighted area in Fig. 2.19c2. We should be concerned with two parts in Fig. 2.20, highlighted by yellow and dark gray boxes. As time evolves from the 199th to 200th steps, Disc gains ground in the yellow part, whereas it loses ground in the dark gray part. This difference is brought about by the difference observed in the *Obs*, payoff, and cooperation-fraction panels.

At the 199th time step, the center agent in the yellow box has Disc & Good. His four (out of five) Disc neighbors happen to be labeled bad. Thus, he defects, while they cooperate. Also, this particular Disc & Good agent has two AllD neighbors who are labelled bad. He defects to those defectors. Therefore, this particular Disc & Good agent happens to exploit his Disc neighbors and is able to successfully avoid being exploited by his AllD neighbors. Thus, he obtains a higher payoff, letting his three AllD neighbors copy Disc from him. The key point in these dynamics is that there is a discriminator labeled good in a Disc-cluster, who has neighboring discriminators labeled bad and defective neighbors labeled bad, which brings a higher payoff to this particular discriminator. To the end, Disc strategy is expanded by his defective agents copying.

Observing the dark gray box at the 199th time step, we notice two Disc & Bad agents facing AllD & Good agents. One important thing to note is that those two Disc & Bad agents are surrounded by discriminators labeled bad. Thus, Disc – Disc relations for those two do not bring them mutual cooperation at all, while Disc – AllD relations impose them to have S. Because of this, those two discriminators on the border suffer from relatively low payoffs and the neighboring defectors obtain relatively high payoffs. Consequently, the two agents convert to defectors at the

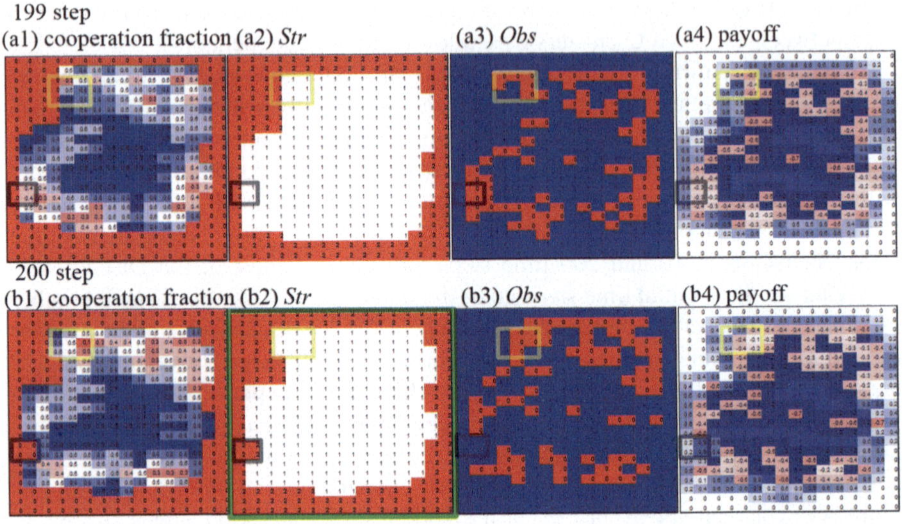

Fig. 2.20 Enlarged snapshots of the area shown by the green-highlighted box in Fig. 2.19c2, which shows what happens at the 199th and 200th time steps. Panels left to right respectively indicate each agent's cooperation fraction, *Str*, *Obs* and payoff per link. Throughout all panels, blue, red, and white indicate cooperative, defective, and neutral. The green-highlighted box in the bottom and center panel indicates that of shown Fig. 2.19c2. See the text for a description of the yellow and dark gray boxes

200th time step. The key point is that there is a discriminator labeled bad in a Disc cluster, having neighboring discriminators also labeled bad and defective neighbors labeled good, which brings a lower payoff to this particular discriminator. To this end, Disc strategy loses ground because of players copying defection.

Those two mechanisms may occur depending upon a local situation that varies from point to point in the domain. An equilibrium level of co-existence would be determined by how frequently those two processes respectively take place as a whole. If the first process takes place more frequently than the second, the global-cooperation fraction would be reasonably high. Contrariwise, if the second process becomes dominant over the first, the global-cooperation fraction would be meagre, even becoming an all-defectors-state under some settings.

The balance between those two opposing processes is determined by the dilemma strength and action-assessment setting.

As observed in Fig. 2.18a, with a sufficiently small dilemma strength, discriminators can be dominant, leading to quite high cooperation. But there would never be an all-cooperators-state, because some discriminators are labeled bad, which degrades the cooperation fraction to some extent. Also, as observed in Fig. 2.18c, presuming a sufficiently large dilemma strength, the second abovementioned process becomes much more frequent, which allows no chance for discriminators to survive. Therefore, the simulation would end with all-defectors-state.

The discussion so far assumes AJ as action assessment. The two other action assessments yield basically the same results as what we see above, e.g., the equilibrium of co-existence is determined by the balance of the two opposing processes; the number of discriminators can either increase or decrease.

When AS is presumed, a mutual-defection relationship is no longer evaluated as good, which is crucially important for suppressing the defectors' prosperity. Inversely, a cooperation–defection relationship in this setting is evaluated as good, unlike AJ. Thus, the discriminator's action in defecting to a neighboring defector, thereby allowing a defector to obtain a low payoff, is not evaluated as good. Thus, this boundary discriminator tends to cooperate, regardless of an opponent's action. This tendency lets his defective neighbor exploit from him. To the end, a discriminator becomes hard to survive by oppressing a neighboring defector vis-à-vis a case under AJ. In nutshell, under the AS assumption, the flux of increasing discriminators is less than that flux of increasing defectors under same dilemma strength as in the AJ case. This explains why we saw a meager cooperation level in Fig. 2.16c as compared with Fig. 2.16a.

When DE is presumed as action assessment, it seems quite natural that we should see a halfway cooperative level (see Fig. 2.16d) of AS and AJ showing, because this action assessment can admire all cooperating actions (mutual cooperation as well as being exploited) like AS and punishing defection (mutual defection) like AJ at the same time. Thus, the flux of increasing discriminators in this setting may be halfway between those under the AS and AJ settings. This qualitatively explains why we could see Fig. 2.16d.

2.3.3.4 High-Resolution Case

Let us explore what happens if we presume maximum resolution for the observed data given to an agent;$Obs \in \{0, 1, 2, 3, 4, 5, 6, 7, 8\}$, which is called the "high-resolution case", as compared with the "low-resolution case," which indicates the default discussed thus far. The agent's strategy is defined by $Str \in \{0, 1, 2, 3, 4, 5, 6, 7, 8, 9\}$, where $Str = 0$ and 9, respectively, refer to the all-defector (AllD) and all-cooperator (AllC) cases. Obs_i, the high-resolution case, directly indicates the number of agent i's actions to their neighbors that are evaluated as good in the previous time step.

Figure 2.21 shows the high-resolution case shown in the same manner as Fig. 2.16, assuming each of the three action assessments.

Interestingly, the high-resolution case, which contains richer information and allows more deliberate strategy, does not always show more enhanced cooperation over the low-resolution case. Obviously, the result presuming AJ (Fig. 2.21a) shows more meagre cooperation than the default case with mixed strategy (Fig. 2.15b). Also, we should note that the AS setting (Fig. 2.21b) shows the best performance among those three, and much better than SPD with a mixed-strategy setting (Fig. 2.15b), which shows a significantly high cooperation level, even at $D_g = D_r = 1$.

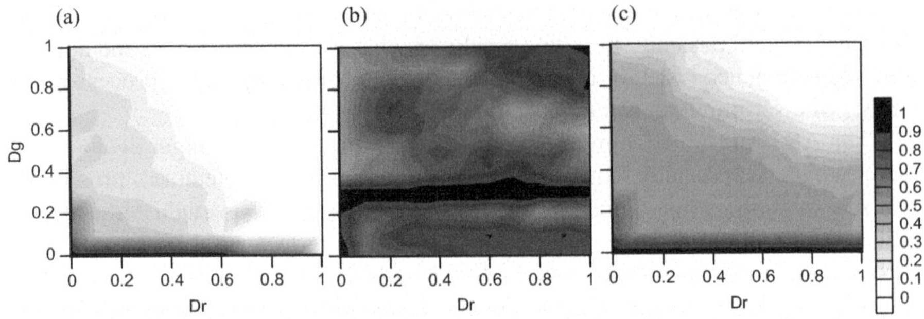

Fig. 2.21 All contours are formatted in the same manner as in Fig. 2.16. Results when high resolution is given; average cooperation fraction presuming (**a**) AJ, (**b**) AS, and (**c**) DE setting

A good question to ask is why we could observe this very specific phenomenon only in the AS setting under high resolution, which even lets agents cooperate in spite of exposing them to a quite large dilemma strength. The necessary condition for this is that the dynamics produce an isolated discriminator's cluster in a sea of bad-labeled defectors (although this detail does not appear in snapshots). The action assessment of AJ and DE can never allow a "sea of bad-labeled defectors", because mutual defection is given a good label by those. This is also prevented in the low-resolution case. Certainly, careful observation of Fig. 2.18 again confirms that only the remaining defectors at equilibrium are not bad, although all are labeled good (see Fig. 2.18b). A bad-labeled defector is not able to enjoy an advantage to exploit neighboring discriminators, because a discriminator will defect to a bad-labeled defector. This substantially explains why the low-resolution cases under the three action assessments do not allow letting dynamics emerge the concerned event. The discussion given here explains why the combination of high resolution and an AS setting is capable of bringing about this specific but interesting event, allowing a high cooperation level even given a large-dilemma situation.

2.3.3.5 Randomness Limit on Obs

We explore what happens if the obtained information is messy. One question is whether our model recovers if the information given through the indirect-reciprocity mechanism becomes meaningless due to a messy signal. As a randomness limit, let us assume an action rule other than Eq. (2.39). Here, we presume Eq. (2.41) as below, where Obs_{random} is a random integer obeying a uniform distribution. It ranges from 0 to 1 in case of low resolution, and from 0 to 8 in case of high resolution.

$$\textit{If } Obs_{random} \geq Str_i \text{ Agent } i \text{ offers Cooperation(C)to Agent } j, \text{ else offers Defection (D)}. \tag{2.41}$$

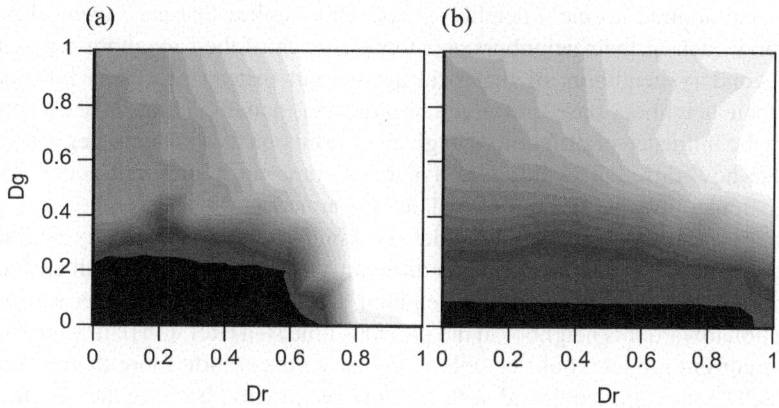

Fig. 2.22 All contours are formatted in the same manner as Fig. 2.15. Social average of cooperation fraction; (**a**) presuming low resolution and Eq. (2.41) instead of Eq. (2.39) is applied; and (**b**) presuming high resolution and Eq. (2.41) is applied instead of Eq. (2.39)

Obviously it makes no difference which action assessment is assumed, whether AS, AJ or DE, because the given information Obs_{random} is mere random noise. One point to be confirmed is that, in the high-resolution case, if an agent's observation of all of his neighbors were to become entirely noisy, Obs_{random} could not be a random integer drawn from the range [0,8] but one drawn from the binominal distribution of ${}_8C_i{\cdot}(1/2)^{8-i}{\cdot}(1-1/2)^i$, where i is an integer in the range [0,8].

Figure 2.22 shows results presuming low and high resolutions.

Because of malfunctioned information process by introducing Eq. (2.41), the model would be equivalent to simple SPD games implementing a multi-discrete-strategy system, where the numbers of discrete strategies are 3 and 9 for low- and high-resolution cases, respectively. In fact, comparing Fig. 2.22 with Fig. 2.15b, we can predict that our model returns to the default SPD model presuming mixed strategy if the given information resolution becomes infinitesimal. It is worthwhile to note that even the case with no meaningful information, Fig. 2.22b, shows better performance than Figs. 2.16c, 2.21a, d. This implies that giving information to agents through an indirect-reciprocity mechanism devastates cooperation which network reciprocity brings as base-line reciprocity, unless appropriate information with presuming apt resolution level and apt action assessment is provided.

2.3.3.6 Summary

Motivated by the fact that spatially distributed agents in the real world may act differently towards their neighbor depending on each neighbor's evaluation, we establish a spatial-prisoner's-dilemma model dovetailed with an indirect-reciprocity mechanism applied to one's neighbors' set. This implies that each agent does not know how each of their neighbors acts towards each of their neighbor's neighbors (i.e., secondary neighbors of the focal agent), but instead processes information about their neighbors based on a certain criterion, namely action assessment. We explore the influence of differences in given information as well as strategy, and also elucidate how different evolutional dynamics come up if different action assessments, stipulating what is good or bad, are presumed.

Extrapolating the result of the model combining network reciprocity with direct reciprocity, in which a focal agent can differently cooperate/defects with each of his neighbors based on each particular neighbor's actions towards him as well as his own action toward this neighbor in the previous time step (Ref. [34]), it was expected that the current model would be able to enhance cooperation more than a conventional SPD game only equipped with network reciprocity, because the added information about an agent's neighbor's attitude should help to establish a robust reciprocity relation with neighbors.

However, our simulation campaign reveals that combining indirect reciprocity frame to network reciprocity as a baseline model is not always meaningful for enhancing cooperation. Also, the fine resolution of both given information and allowed strategy is not always meaningful for obtaining more cooperation as compared with presuming low resolution.

In the low-resolution case, the action assessment that admires cooperators who only cooperate with other cooperators and defectors who only defect to other defectors shows significantly enhanced cooperation. However, this action assessment does not work well in the high-resolution case.

In the high-resolution case, however, an action assessment that seems simpler than above, namely admiring cooperators, outperforms other action assessments. In particular, this setting shows relatively high cooperation even in stronger-dilemma games.

In the present study, we considered no mutation effect when an agent updates his strategy. From a realistic applications point of view, noise effects resulting from both action and copy errors must be taken account, which is an interesting question for future work.

We readily recognize that what is shown here just relies on an inductive approach, namely a simulation procedure with presuming specific premises, and does not rely on a deductive, or say, theoretical approach, which certainly makes it difficult to capture a more universal view and to draw a more general conclusion. But we daresay it is still meaningful, because our finding successfully highlights that combining network reciprocity with indirect reciprocity may result in negative interference to enhance cooperation. In this way, a simulation approach should work to seek any other new interesting points so as to provide a guiding light for searching for a next point to dig by means of an analytical approach.

2.4 Noise-Driven Network Reciprocity

The term of "stochastic resonance" is one of the hottest topics in the field of statistical physics. In its usual context, noise is recognized as unwanted gain that should be removed. But in recent decades, quite a few physicists have found an interesting phenomenon called the stochastic-resonance effect, whereby adding some appropriate noise to a non-linear system featuring some cyclicity enhances the rhythm laying behind, which somehow brings useful effects.[44]

Some evolutionary game theorists have been attracted the stochastic resonance, since they believe that additional noise may enhance cooperation significantly. In fact, there have been many precursors who have tried to discover a stochastic-resonance effect in SPD games. Spatial-game settings rely on network reciprocity, which has two important functions for lessening anonymity (i.e., adding social viscosity) among players: limiting the number of game opponents, and adding a local-adaptation mechanism, by which an agent copies a strategy from a neighbor linked by a network. These explain how cooperators can survive in a system with social dilemmas, even though agents are required to use only simple strategies of cooperation (C) or defection (D).

Meanwhile in SPD assuming a finite population, a demographic fluctuation can be observed intrinsically from a system-dynamics point of view. In view of statistical physics, such dynamical systems may show a stochastic resonance if exposed to an appropriate noise that is extrinsically given through an additional mechanism, where cooperation, originally vanishing, can survive or even surge compared to defection. Along with this context, there have been many works concerned with what noise (and by how) can be additionally imposed on the SPD model to effectively enhance network reciprocity.

Two important subordinates significantly affect the final level of network reciprocity in an SPD model: an underlying network and a strategy-updating rule. In opposition to homogeneous networks, presuming a heterogeneous topology may bring some noise-driven resonance, as Perc carefully explored.[45] With respect to an update rule, some pioneers investigated how the noise coefficient in the Fermi function κ (explained later) can bring enhanced cooperation through stochastic resonance in various situations.[46,47,48,49,50] This noise coefficient controls the extent of how a certain updating process becomes either deterministic ($\kappa \to 0$) or perfectly random ($\kappa \to \infty$). We exclude, in the following discussion, those two aspects resulting from heterogeneity of the underlying network and randomness brought

[44]Benzi et al. (1981).

[45]Perc (2006a).

[46]Perc and Marhl (2006).

[47]Chen and Wang (2010).

[48]Szolnoki et al. (2009a).

[49]Du and Fu (2013).

[50]Szolnoki et al. (2009b).

about by the updating rule from what we call the "noise" effect, because these aspects obviously comprise a very fundamental base of the SPD model, which should be regarded not as "additional", but rather indispensable. Instead, in our numerical exploration, we will vary the network between the lattice and scale-free graph as respectively representing homogeneous and heterogeneous topology, and vary the update rule between Imitation Max and Pair-wise Fermi (both explained later), which respectively represent deterministic and stochastic rules.

Concerning the addition of noise to an SPD model, two major concepts have been considered: action error and copy error. Action error,[51,52] presumes an erroneous action taken by an agent, regardless of their strategy during a gaming process, which inputs noise to the original dynamics. Meanwhile, copy error[53,54,55] presumes a situation where either an agent miss-copies from their neighbors or where their strategy is randomly mutated. There is another concept, termed observational error,[56,57] which is relevant to copy error. In a strategy-updating process, a focal agent evaluates his own payoff and those of any one of his neighbors or of all of his neighbors. The observation error means mis-accounting for those evaluations, which inevitably leads to a malfunction in the copying process.

Meanwhile, Perc[58] originated a concept called the payoff-noise model, which was expanded by other researchers.[59,60,61,62,63,64] According to this model, either the payoff matrix or the accumulated payoff after gaming is biased by an additive noise whose average is kept at zero. Those previous works found more enhanced network reciprocity than the default SPD model. Obviously observation error and the payoff-noise model share the same base, although the details may differ.

As another noisy situation, we can suppose the following. In a network there are a certain fraction of agents who behave randomly and never update strategy; we should call these noisy and random agents. This idea is somehow analogous to what Masuda called "zealots" [65] which presumes that there are stubborn cooperators in the network who never defect and never update strategy. Unlike his work, in the

[51] Helbing and Yu (2009).
[52] EL-Seidy (2015).
[53] Brede (2013).
[54] Yokoi et al. (2014).
[55] Gao et al. (2015).
[56] Hao et al. (2015).
[57] Zhang et al. (2016).
[58] Perc (2006b).
[59] Tanimoto (2007).
[60] Zhang et al. (2013).
[61] Yao and Chen (2014).
[62] Hauang et al. (2015).
[63] Perc (2007).
[64] Shen et al. (2016).
[65] Masuda (2012).

present study, we should presume same the number of zealous cooperators and zealous defectors or random-action agents in terms of equitability between C and D.

Thus, based on the literature, we should explore action error, copy error, observational error, random agents, and zealous agents when we refer to noise effects in SPD games.

2.4.1 Model Setup

We presume a standard SPD-game setting. $N = 10^4$ agents are placed on each of the vertices of the assumed underlying network explained below. Each agent plays a PD with all his neighbors and accumulates payoffs resulting from all such games. In a game, a player receives a reward (*R*) for mutual cooperation and a punishment (*P*) for mutual defection. If one player chooses cooperation (C) and the other defection (D), the latter obtains a temptation payoff (*T*), and the former the sucker's payoff (*S*). We assume a spatial game with $R = 1$ and $P = 0$, parametrized as $\begin{pmatrix} R & S \\ T & P \end{pmatrix} = \begin{pmatrix} 1 & -D_r \\ 1 + D_g & 0 \end{pmatrix}$, where $D_g = T - R$ and $D_r = P - S$ imply a chicken-type dilemma and stag-hunt dilemma, respectively. We limit the PD game class by assuming $0 \le D_g \le 1$ and $0 \le D_r \le 1$.

We vary the strategy-updating rule between Imitation Max (IM) and Pair-wise Fermi (PW-Fermi). IM is the best-accepted deterministic update rule, wherein a focal player copies the strategy of the player among his neighbors and himself who earns the largest payoff during the current time step. Also, we presumed PW-Fermi to be the most representative stochastic-update rule, where a player compares his accumulated payoff (Π_i) with that of a randomly selected neighbor (Π_j) and copies the neighbor's strategy according to $P_{copy}^{i \leftarrow j} = \frac{1}{1 + \exp\left[\left(\Pi_i - \Pi_j\right)/\kappa\right]}$. Here, κ indicates noise coefficient, which is presumed to be 0.1 throughout this study.

As a population structure, we assume a 2D lattice (hereafter, Lattice) with a degree of 8 ($k = 8$), i.e., a Moore neighborhood and scale-free network by Barabasi–Albert algorithm[66] (hereafter, BA-SF).

Under each simulation setting, we explore how each of the different error settings presumed below influence network reciprocity by varying whether updating is deterministic or stochastic, {IM, PW-Fermi}, and whether the underlying network is homogeneous or heterogeneous, {Lattice, BA-SF}.

We presumed eight different models of error setting and one combined model as follows:

Action error #1: With a probability of ε_{a1}, an agent randomly cooperates or defects regardless of his strategy. We presume; $\varepsilon_{a1} \in \{0.05, 0.10, 0.20\}$.

---

[66] Barabasi and Albert (1999).

Action error #2: With a probability of ε_{a2}, a cooperator defects and a defector cooperates. This means a limited mixed-strategy setting, whereby a cooperator (defector) cooperates with a probability of 1 - ε_{A2} (ε_{A2}). We presume; $\varepsilon_{a2} \in \{0.05, 0.10, 0.20\}$.

Copy error #1: This error emulates the situation of a random mutation taking place in a copying process. With a probability of ε_{c1}, an agent fails the fair-copying process, and becomes either C or D with equal probability. We presume; $\varepsilon_{c1} \in \{0.05, 0.10, 0.20\}$.

Copy error #2: This error emulates a "mishandling" event taking place in the copying process, which is influenced by the formation of one's neighbors. In the case of IM, with a probability of ε_{c2}, an agent randomly selects one of his neighbors to copy irrespective to the selected neighbor's payoff. In the case of PW-Fermi, with a probability of ε_{c2}, an agent either keeps his strategy or copies the strategy of a random neighbor. Either option has an equal probability. We presume; $\varepsilon_{c2} \in \{0.05, 0.10, 0.20\}$.

Observation error #1: During a strategy update, an agent independently misobserves each of all accumulated payoffs of both himself and his neighbors by $\delta_{ob1} \cdot (R - P) \cdot Rnd[] \cdot k_j$, where $-1 \leq Rnd[] \leq 1$ indicates a real number obeying a uniform distribution and k_j indicates agent j's degree. This is expected to bring about an analogous effect to Tanimoto[59] and Perc[58]'s so-called payoff-noise model. We presume; $\delta_{a1} \epsilon \{0.25, 0.50\}$.

Observation error #2: Referring to the previous setting, this presumes only δ_{ob2} of all agents miss-observe each of the accumulated payoffs of both himself and his neighbors by $(R - P) \cdot Rnd[] \cdot k_j$. We presume; $\delta_{a2} \epsilon \{0.10, 0.25\}$.

Random agents: With a fraction of ε_{ra}, there are agents who behave randomly as C or D and never update their strategy. This resembles the so-called zealous model[67] where agents always cooperate regardless of strategy updating. This can be said to be a specific variant of Action error #1. With Random agents, agents cooperating or defecting at random are specially frozen, unlike for Action error #1. We presume; $\varepsilon_{ra} \in \{0.05, 0.10, 0.20\}$.

Zealous agents; With a fraction of $\varepsilon_{za}/2$, there are respectively stubborn cooperators and defectors who never update their strategy. This is analogous to the zealous model[67] but differs because of the existence of zealous defectors as well as zealous cooperators.[68] We presume; $\varepsilon_{za} \in \{0.05, 0.10, 0.20\}$.

Action & Copy errors: It presumes both Action error #1 and Copy error #2 simultaneously.

We assume synchronous updating; thus, the strategy updating for the entire population follows after the all agents gaming as shown in Fig. 2.23. Action error takes place in the gaming process, while copy and observation errors happen in the strategy-updating process at each time step of the procedure. The settings of random

[67]Matsuzawa et al. (2016).

[68]Fukuda and Tanimoto (2016).

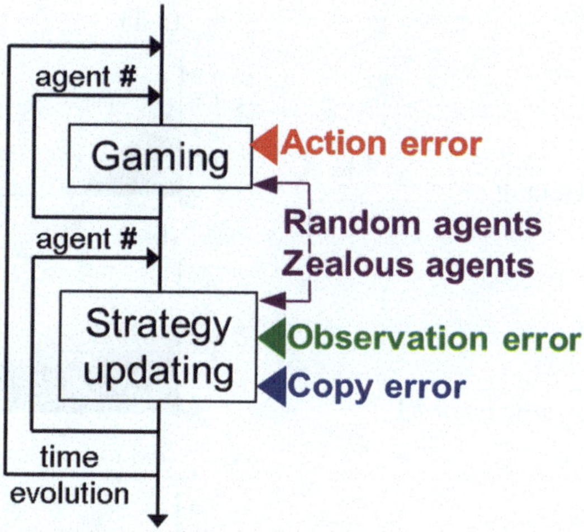

Fig. 2.23 When each presumed error setting taking place in the flow of SPD games is shown

and zealous agents relate to both gaming and strategy updating, because those who randomly or zealously behave in a game and are excluded from the strategy-updating process.

Each simulation episode was performed as follows. Beginning from an initial arrangement in which an equal number of cooperators and defectors were randomly distributed, several simulation time-steps were run until the fraction of cooperators reached quasi-equilibrium (which can be approximately regarded as an equilibrium of the numerical process). If the global-cooperation fraction, P_c, continues to fluctuate, we use the average fraction of cooperation over the last 250 time-steps of a 10^4-time-step run. We vary the dilemma strength to cover PD, $0 \leq D_g \leq 1$ and $0 \leq D_r \leq 1$, where the increment interval is 0.1. The results shown below were drawn from 100 runs—that is, each ensemble average was formed from 100 independent simulations.

2.4.2 *Synopsis Result*

Figure 2.24 shows the averaged cooperation fraction of 100 realizations ranging over $0 \leq D_g \leq 1$ and $0 \leq D_r \leq 1$ for each error setting defined in Sect. 2.2 when presuming IM and Lattice. For comparison, the default setting (panel (a-1)) without any noise (i.e., the simple network reciprocity resulting from a combination of IM + Lattice) is also given. Panels (b-1), (b-2) and (b-3) indicate the results of action error #1 with ε_{a1}=0.05, 0.10, and 0.20. Figures 2.3, 2.4, and 2.5 respectively show the results for counterparts assuming IM + BA-SF, PW-Fermi + Lattice, and PW-Fermi + BA-SF.

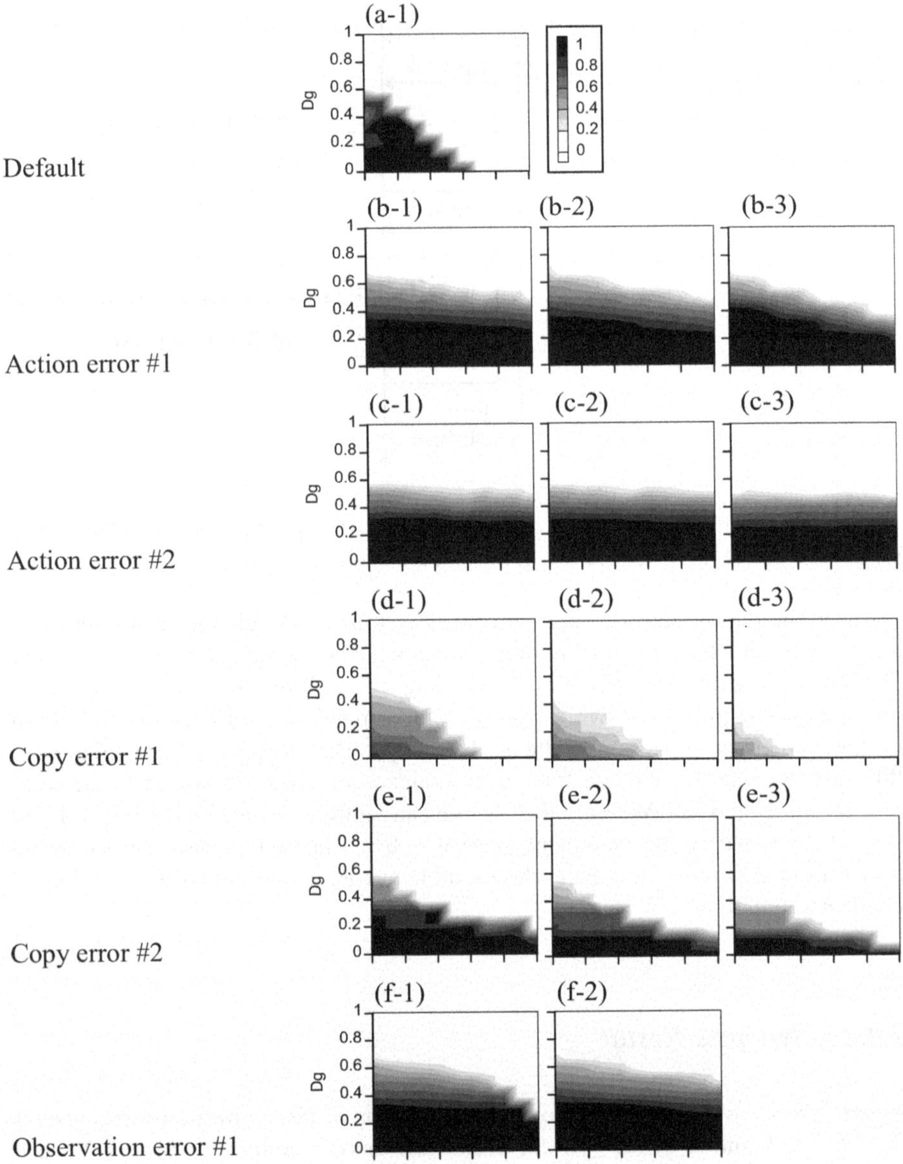

Fig. 2.24 Results presuming an IM + Lattice. Average cooperation fraction for 100 realizations

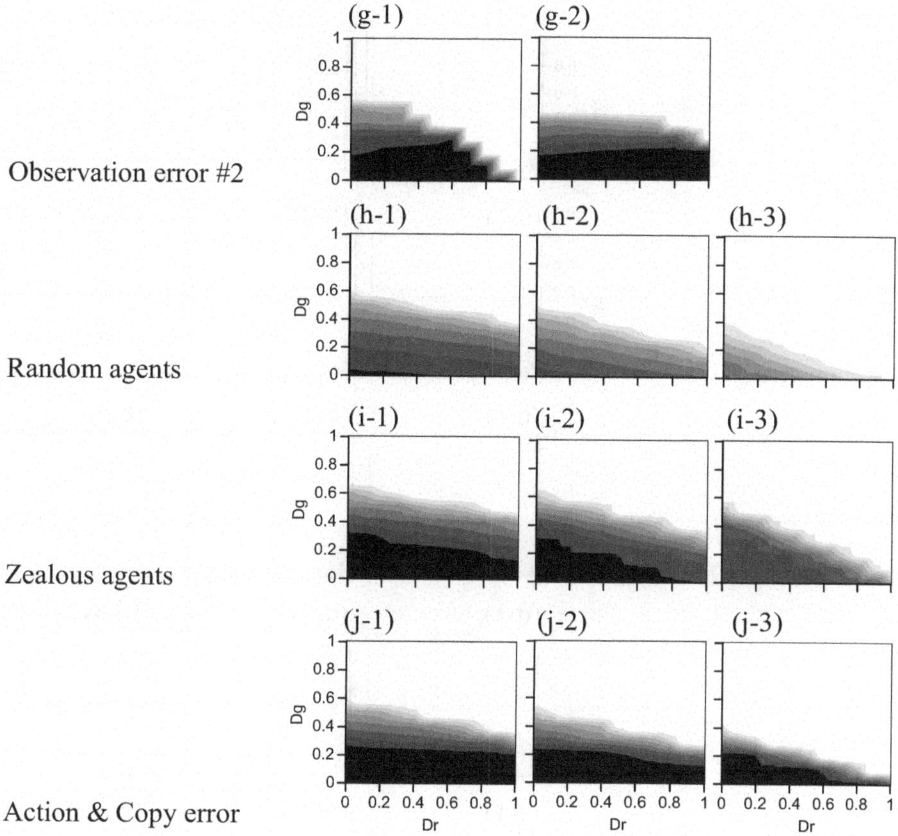

Fig. 2.24 (continued)

Viewing Figs. 2.25, 2.26, and 2.27, what we should confirm before starting a detailed discussion is that, even the noise-driven effect alone would differ significantly depending on of the type of noise and other settings, as well as the underlying network and update rule. Although some cases can show network reciprocity beyond the original case, others can have a worse influence, indicating that the original network reciprocity can be devastated.

As a general tendency, more enhanced network reciprocity can be observed in the two cases presuming IM as update rule than each of the default settings, but less when presuming PW-Fermi. This implies that further enhancement by adding noise might be possible under a deterministic-update rule like IM but less so under a stochastic one like PW-Fermi. It is likely that a too-excessive noise over the base of PW-Fermi, instinctively containing a noise effect brought about by the Fermi function, deteriorates the original network reciprocity rather than bolstering it.

Let us compare each default setting with the various assumed noise settings as a whole.

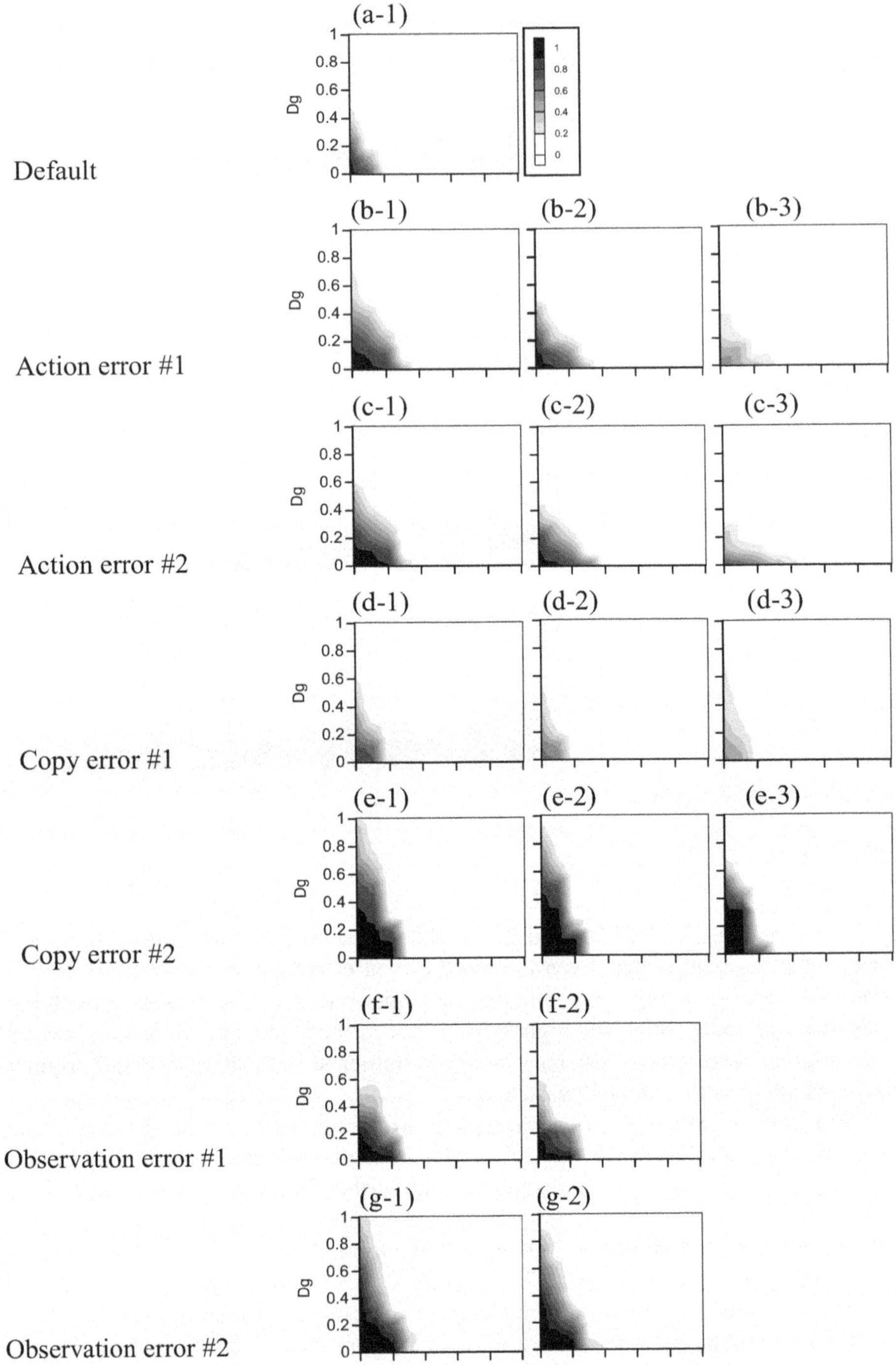

Fig. 2.25 Results presuming IM + BA-SF. Average cooperation fraction for 100 realizations

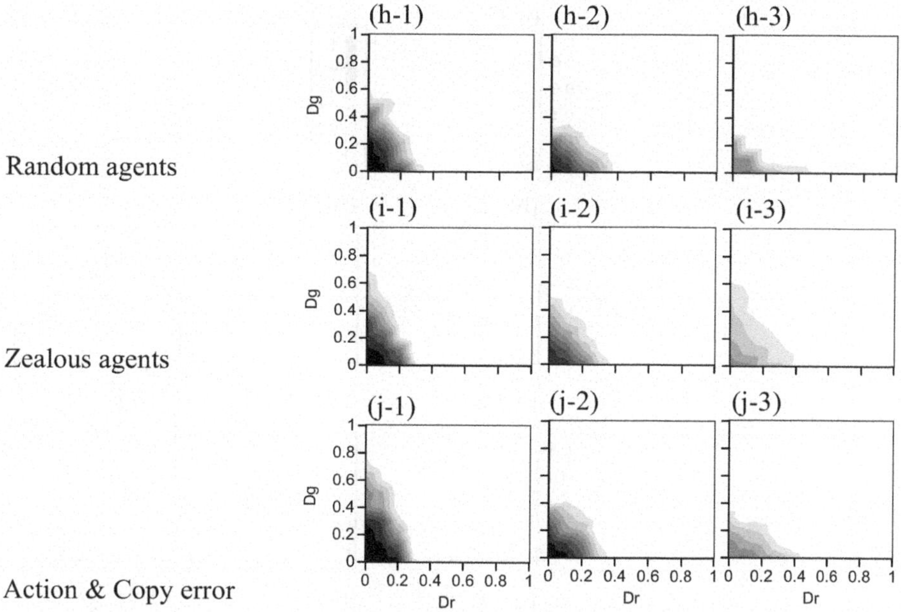

Fig. 2.25 (continued)

In the case of IM + Lattice, overall, adding a noise enhances cooperation in games where an SH-type dilemma predominates over a Chicken-type dilemma. In the case of IM + BA-SF, a noise enhances cooperation by surmounting both dilemmas.

When presuming PW-Fermi + Lattice, the original network reciprocity is so small that noise influence seems too subtle to be observed. In the case of PW-Fermi + BA-SF, the default network reciprocity is larger than that in the case of IM + BA-SF. However, unlike in IM + BA-SF, adding a noise does not show an obvious tendency of enhancing cooperation.

2.4.3 *In-Depth Discussion*

Regardless of the presumed combination of update rule and underlying network, Copy error #1 deteriorates network reciprocity, showing worse cooperation than each of the default cases. Obviously, unlike other noise settings, Copy error #1 lightens the color in each contour, indicating that a relatively high cooperation cannot be attained at equilibrium. This is because, under any updating rule or network topology, the copy error that randomly chooses either C or D rather than using an appropriate strategy causes the adaptation process, which is crucially important for realizing instinct network reciprocity, to malfunction.

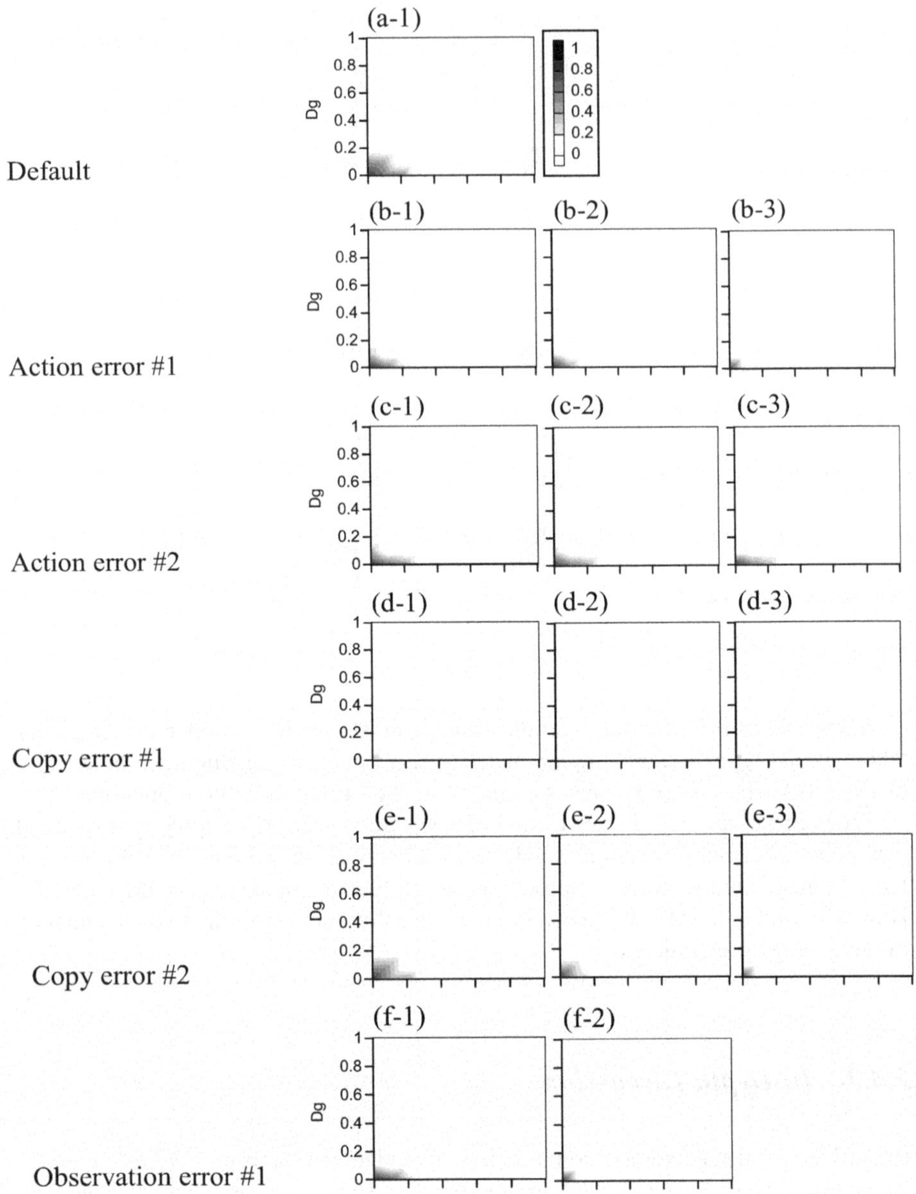

Fig. 2.26 Results presuming an PW-Fermi + Lattice. Average cooperation fraction for 100 realizations

Observation error #2

Random agents

Zealous agents

Action & Copy error

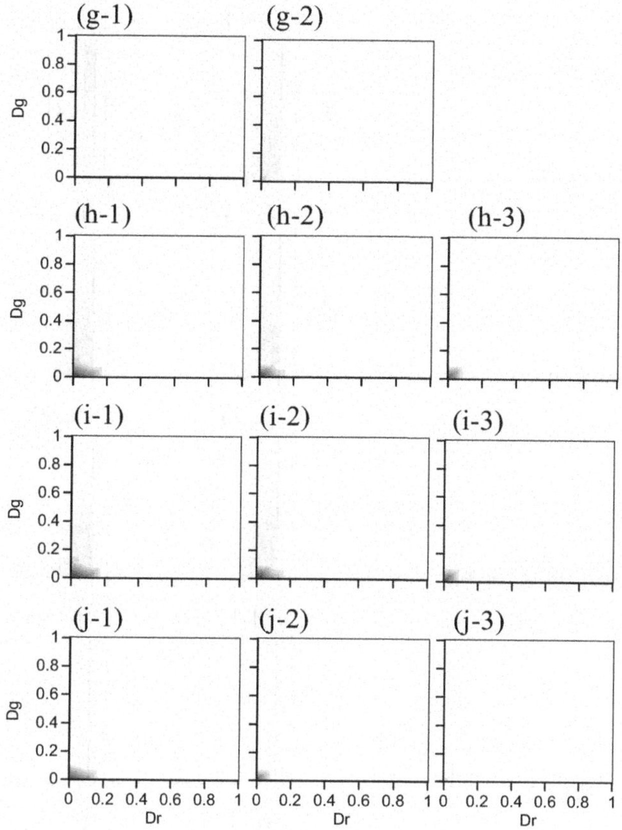

Fig. 2.26 (continued)

Comparing Action error #1 and Random agents, all results with the same error rate seem almost same except for IM + Lattice. As mentioned in the model setup, the only different between Random agents and Action error #1 in the presumed process is that the spatial position of each of agents cooperating or defecting with equal probability is initially assigned, and since then frozen in an episode unlike Action error #1 presuming. Thus, bringing almost same process to give a noise effect in those two settings, all cases other than IM + Lattice show almost same tendency. Since the IM + Lattice contains no stochastic processes in either updating or the underlying network, giving random noise to agent's actions, even if such actions would be entirely cooperative, deteriorates cooperation. This can be qualitatively justified by referring to the study of zealous cooperators by Matsuzawa et al.,[67] where it was found that placing zealous cooperators under the IM + Lattice setting devastates cooperation compared to network reciprocity without zealous cooperators. This is because the dynamics in the EXP period (see Ref [46], and Fig. 2.28) is badly influenced by existing agents who never do with strategy updating when a cooperative cluster

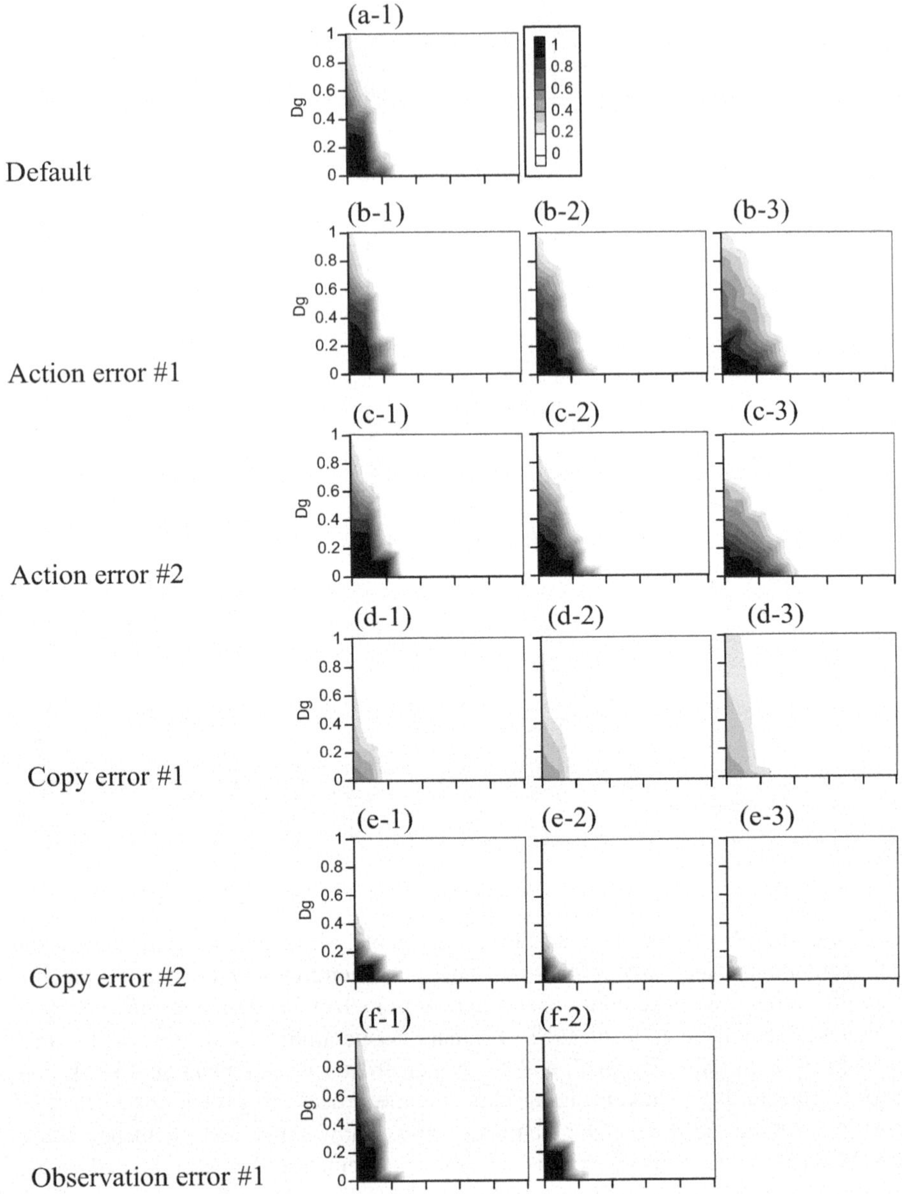

Fig. 2.27 Results presuming PW-Fermi + BA-SF. Average cooperation fraction for 100 realizations

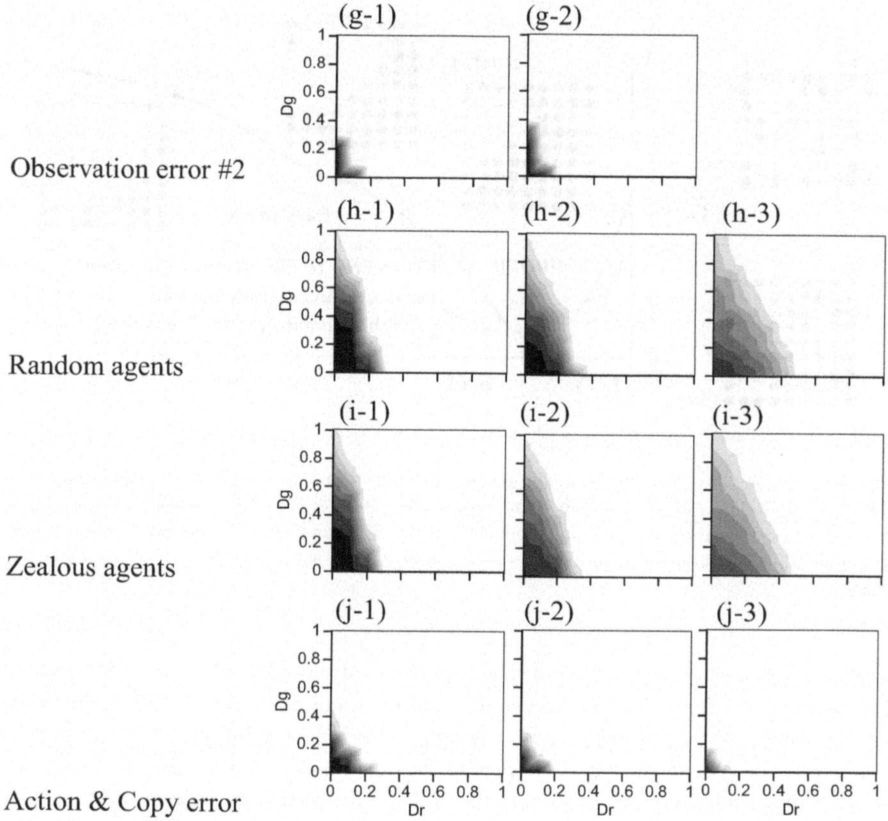

Fig. 2.27 (continued)

surviving the END period starts to expand into the neighboring domain occupied by defectors. If each of those random agents were not fixed at a certain position in the domain, this influence would be less devastating for instinct network reciprocity (this is the case of action error #1). However, if random agents are spatially fixed (as in the random-agent case), meagre network reciprocity is the result.

For the IM + Lattice, all noise settings except for Copy error #1 generally show enhanced cooperation compared to the default case in the region where D_r is much larger than D_g. This can be explained in terms of END and EXP. The default setting of IM + Lattice without any noise is deterministic except for having a stochastic event of initial spatial assignment of an equal number of cooperators and defectors in the domain. Thus, if a perfect C-cluster with a block of nine ($= 3 \times 3$) cooperators happens to be configurated in the initial assignment, it never perishes as long as $D_g < \frac{5}{3}$. A perfect C-cluster can expand to the neighboring defectors' domain as long as $D_g + D_r < \frac{2}{3}$. The triangular region of high cooperation is observed under the default setting, which is consistent with the region of $D_g + D_r < \frac{2}{3}$. This fact implies that if an initial configuration contains some perfect C-clusters, cooperation does not

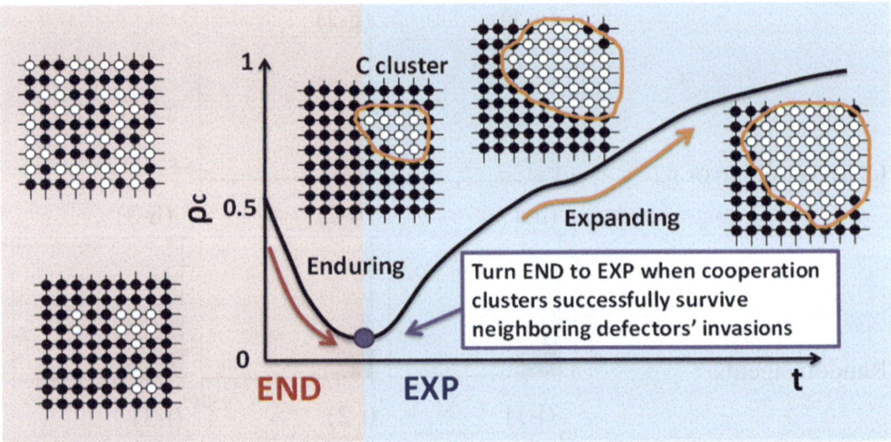

Fig. 2.28 Schematic view for the evolution of cooperation in a spatial prisoner's dilemma game. *Enduring (END) period*: Initial cooperators will be rapidly overtaken by defectors, causing only a few cooperators to be left by forming compact C-clusters. *Expanding (EXP) period*: C-clusters start to expand, since a cooperator on the clusters' border can attract a neighboring defector into the cluster

grow as long as $D_g + D_r > \frac{2}{3}$, i.e., cooperators almost perish except for perfect C-clusters (see Fig. 2.29). However, by adding a certain type of noise other than copy error #1, $D_g < D_r$ enables not only a perfect C-cluster, but also other imperfect C-clusters, to survive in the END period, because a defector is less advantageous over neighboring cooperators than in the $D_g \approx D_r$ situation. Moreover, after surviving the END period, those C-clusters may be able to expand in the EXP period to the neighboring defective space, in spite of $D_g + D_r > \frac{2}{3}$. This is because a noise incurs an irregular expansion of cooperators, even where cooperation is quantified as being inferior to defection, although an inverse event (a cooperator dying even under $D_g + D_r < \frac{2}{3}$) may stochastically happen due to the noise effect. For example, let us consider Fig. 2.30, where random agents with settings $D_g = 0.3$ and $D_r = 0.8$ help C-clusters to survive the END and expand in the EXP period, as discussed above.

Comparing the random- and zealous-agents settings, the latter results in a better cooperation fraction than the former for IM + Lattice. A less significant version of the same tendency can be observed in the IM + BA-SF case. In the PW-Fermi + Lattice and PW-Fermi + BA-SF cases, there seems no difference between the two settings.

Figure 2.31 compares the default setting presuming an IM + Lattice with the result when presuming a mixed strategy instead of a binary C or D strategy (hereafter: discrete-strategy system), where an agent has a real strategy value ranging from 0 to 1 that indicates a probability of cooperating or defecting. Again, those two presume the IM + Lattice case. As described in the model setup, Action error #2 partially contains a mixed-strategy system, although it relies on a discrete-strategy system as its base. More precisely, it can be called a binary-strategy system where

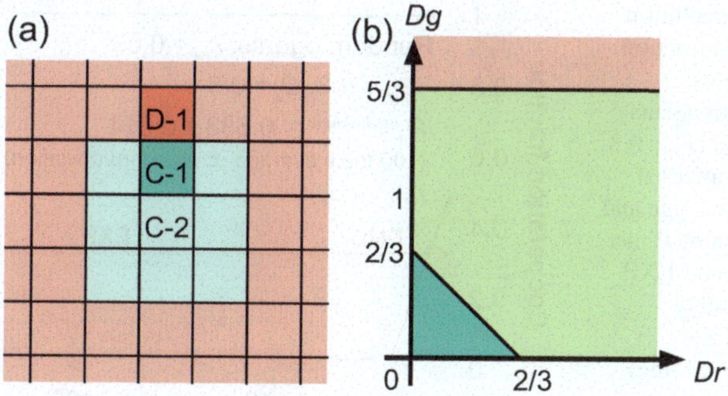

Fig. 2.29 Conditions under which a perfect C-cluster can survive and expand in the case where a lattice and IM are assumed. Let us presume a perfect C-cluster in a sea of defectors, as shown in Panel (**a**). Agent (D-1) neighboring the perfect C-cluster most effectively exploits the neighboring cooperator, who gets 3 T + 5P. This is re-written as 3 · (1 + D_g) when substituted into $\begin{pmatrix} R & S \\ T & P \end{pmatrix} = \begin{pmatrix} 1 & -D_r \\ 1+D_g & 0 \end{pmatrix}$. His neighbor, cooperative agent (C-1), is exploited by three neighboring defectors, and thus only earns $5R + 3S = 5 - 3 \cdot D_r$. However, one of his neighbors, agent (C-2), at the center of the perfect C-cluster, gains a high payoff, $8R = 8$. Even if an agent is severely exploited by his defective neighbors, the IM rule compels him to keep cooperating as long as he has a cooperative neighbor who obtains a high payoff. For these reasons, we can infer that a perfect C-cluster initially placed at the center of an all-defectors domain never perishes as long as $D_g < \frac{5}{3}$. We also infer that this perfect C-cluster can expand in the domain as long as $D_g + D_r < \frac{2}{3}$. These two particular conditions are shown in Panel (**b**), in which the green and blue regions meet at the first inequality, and the blue region represents where (D_g, D_r) satisfies the second inequality. Therefore, we can deduce that a dynamical episode starting with a single perfect C-cluster in a sea of defectors and satisfying the second inequality could ultimately attain an all-cooperators-state if we assume a sufficiently large domain. Paraphrasing this, as long as $0 \le D_g \le 1$ and $0 \le D_r \le 1$ are satisfied, any evolutionary path implemented with the typical network-reciprocity mechanism can never be absorbed by an all-defectors-state, and that at least the initial nine cooperators will always remain, implying that there is actually no END period. Instead, this particular initial setting ensures that every evolutionary path starts immediately in the EXP period, not undergoing any END period as cooperation can grow immediately from this initial state

two mixed strategies, $p_1 = 1 - \varepsilon_{ae2}$ and $p_2 = \varepsilon_{ae2}$, are presumed. Comparing Fig. 2.31b with the result presuming IM + Lattice and Action error #2, the performance enhancement realized by each in the small-D_g region (lower half of both panels) is almost the same, although the full mixed-strategy system (Fig. 2.31b) shows more enhanced cooperation than does Action error #2 in the large-D_g region. With increasing D_g, it would be hard to survive a relative cooperator ($p_1 = 1 - \varepsilon_{ae2}$) as opposed to a relative defector ($p_2 = \varepsilon_{ae2}$). As Nowak & May[18] reported, a Chicken-type dilemma with spatial settings disfavors forming massive C-clusters as in usual SPD games, but rather favors forming string-like C-clusters, which implies that, as long as a traditional discrete-strategy system is relied upon, the network reciprocity for solving PD games becomes less effective with increasing D_g. Instead of a discrete-strategy system, a mixed-strategy system may help to solve this drawback, because a mutual offering of either C or D based on the strategy p denoted

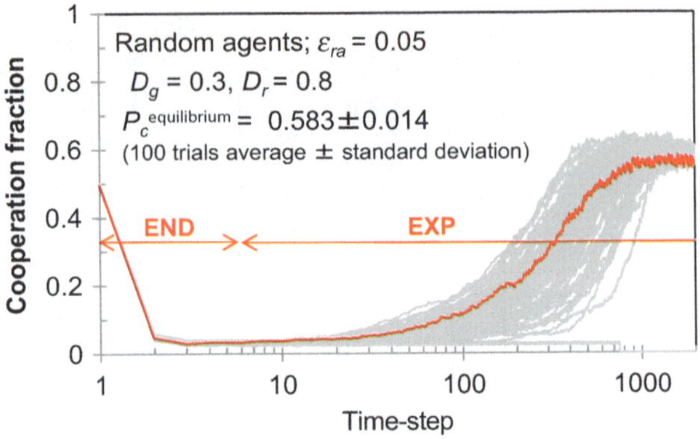

Fig. 2.30 Time evolution of the cooperation fraction (P_c) in all 100 cases, presuming random agents with $D_g = 0.3$ and $D_r = 0.8$, $\varepsilon_{ra} = 0.05$. As numerical information, the average and standard deviation of P_c are given. The END and EXP periods are separately shown for one of the 100 realizations (red line)

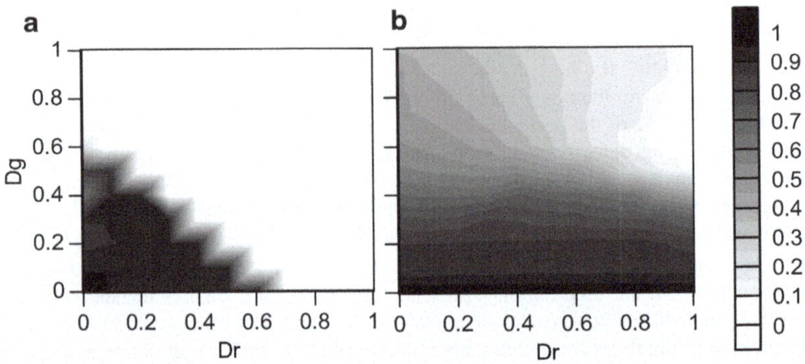

Fig. 2.31 Average cooperation fraction of 100 realizations presuming IM + Lattice. Panel (**a**) is the default setting based on a discrete-strategy system where either C or D can be offered. This is copied from Fig. 2.2a. Panel (**b**) gives the result based on a mixed-strategy system in which an agent stochastically offers either C or D based on his strategy, which is defined as a real value ranging from 0 to 1

by a real number approximately realizes an internal equilibrium appearing in real chicken games.[69,70] Confirming all of those points, the comparing two results above-mentioned might be understood as likely.

Let us cast a spotlight upon Observation errors #1 and #2 in both IM + Lattice and IM + BA-SF. In IM + BA-SF, Observation error #2 shows more enhanced cooperation than Observation error #1. However, for IM + Lattice, Observation error #1 is superior to #2. Observation error #1 presumes that all agents mis-observe, while #2 presumes that only selected agents mis-observe but with a much greater assumed error deviation. This difference from Observation error #1, concentrated noise input giving into limited agents, could help Observation error #2 to push enhanced

[69]Zhong et al. (2012).

[70]Kokubo et al. (2015).

cooperation up more than Observation error #1 does if some of the "limited agents" happen to be so-called hub-agents. Thus, the setting of Observation error #2 effectively works with a heterogeneous network like BA-SF.

The case of duplicated action and copy errors implies that a superimposed-noise effect cannot be expected; rather, the error effect is approximately the same as the lower of Action error #1 and Copy error #2. Coming to think the fact that a SPD game process consisting of gaming and strategy updating is typically non-linear, it would be likely.

2.4.4 *Summary*

We started this discussion of the term "stochastic-resonance effect", which implies that noise is not annoying but may rather bring some willing result to us. Pushed by the respectably rich supply of previous works on SPD games and motivated by the fact that the question of how different noise mechanisms can bring about different noise-driven network reciprocities has not been answered, we explored a series of simulations wherein we compared action, copy, and observation error under placement of random or biased agents or variants thereof. The effect significantly differs depending on the error setting, as well as the presumed underlying topology and strategy-updating rule. In general, the action error shows better performance for enhancing cooperation than copy error. As long as Imitation Max is presumed as an update rule, observation error, which shares its basic concept with the "payoff-noise model", can show a relatively high performance for enhancing cooperation.

In the present study, we have only dealt with the static noise, which means that the presumed framework for adding noise is not altered by time evolution. One of our next works concerns dynamic noise, by which a noise framework may change according to the stage in time evolution.[71] The question of how different dynamic noises based on different noise mechanisms bring different network reciprocities may constitute an interesting challenge.

2.5 Do Sanctions Triggered by Jealousy Help Grow Cooperation?

As we have repeatedly discussed, in human society, cooperation is crucial for maintaining public welfare. Per the Darwinian theory of evolution, the principles of human behavior are governed by competition rather than cooperation. The fact that cooperation is universally observed across cultures has therefore resulted in one long-running challenge in evolutionary biology and social science and question of why humans cooperate with unrelated individuals. One framework that can be used to

[71] Tanimoto and Ogasawara (2015).

justify the promotion of cooperation is costly punishment.[72,73,74,75,76,77,78,79] Such punishment, by which an agent can pay a cost to impose a fine on defective agents, comes in two forms: individually inspired sanctions through face-to-face interactions (peer punishment[80,81,82,83,84]) and institutional sanctions (pool punishments[85,86,87]). In both cases, penalizing defectors reduces the likelihood of defection. However, the punishment involves a cost burden for the punishing agents as well as the defector. Consequently, it inevitably diminishes the overall social payoff. Therefore, when considering possible punishment frameworks, both the evolution of cooperation and the social efficiency of such frameworks should be discussed.[88,89,90,91]

2.5.1 Model Setup

Once again, we consider SPD games. Each agent adopts either a strategy of cooperation (C) or defection (D) and receives a payoff through an interaction. The payoff is determined by a combination of the strategies of both agents and can be defined as punishment (P), reward (R), sucker (S), and temptation (T) for the D-D, C-C, C-D, and D-C interactions, respectively. Following our previous study Ref [3], [8], let us introduce the dilemma parameters, the gamble-intending dilemma, $D_g = T\text{-}R$, and the risk-averting dilemma, $D_r = P\text{-}S$, presuming $R = 1$ and $P = 0$ without loss of generality. Thus, the payoff matrix can be denoted as $\begin{pmatrix} R & S \\ T & P \end{pmatrix} = \begin{pmatrix} 1 & -D_r \\ 1 + D_g & 0 \end{pmatrix}$, where we limit $D_g > 0$ and $D_r > 0$, resulting in

72Sigmund et al. (2001).
73Brandt et al. (2003).
74Nakamaru and Iwasa (2005).
75Helbing et al. (2010).
76Boyd et al. (2010).
77Baldassarri and Grossman (2011).
78Sasaki et al. (2012).
79Cui and Wu (2014).
80Fehr and Gächter (2002).
81Gardner and West (2004).
82Brandt et al. (2006).
83Gächter et al. (2008).
84Rockenbach and Milinski (2009).
85Gurerk et al. (2006).
86Sigmund et al. (2010).
87Szolnoki et al. (2011a).
88Fehr and Fischbacher (2003).
89Fowler (2005).
90Milinski and Rockenbach (2008).
91Dreber et al. (2008).

a game that belongs to the PD class. In the following discussion, we impose the additional limits $0 \leq D_g \leq 1$ and $0 \leq D_r \leq 1$.

The total number of agents is set to $N = 10^4$. We adopt two spatial-population structures: a square lattice with a Moore neighborhood (degree of $k = 8$) and a scale-free network following the Barabási–Albert model (hereafter BA-SF) with $<k> = 8$.

For each simulation episode, equal numbers of Cs and Ds are randomly distributed on an underlying network as an initial condition. Each agent plays games with their neighbors on the network and accumulates a payoff at each time step. An agent decides whether to introduce a jealous sanction based on an evaluation of the payoff situation between himself and each of his neighbors. If his opponent's payoff, π_j, is larger than his own (π_i) but less than twice of its value ($\pi_i < \pi_j < 2\pi_i$), the focal agent enforces a sanction with a payoff-based probability (π_i - π_j)/π_i. In this case, the agent pays a sanction cost that is some fraction of his own accumulated payoff, c_s, so as to introduce a sanction on his opponent in order to reduce the opponent's payoff, p, to some extent. It might be justified to assume that there is an upper limit on the jealous sanction, since one tends to feel less jealousy directed toward people who are far richer. To discuss the efficiency of the sanction described above, let us introduce two parameters: a cost coefficient, c_s, and a penalty coefficient, p, as follows:

$$\begin{aligned} \pi'_i &= (1 - c_s)\pi_i \\ \pi'_j &= \pi_j - p\pi_i. \end{aligned} \tag{2.42}$$

Here, we define the efficiency of a sanction via a cost-penalty ratio, $\beta = p/c_s$. Eq. (2.42) implies that even if a sanction enforcer reduces a player's own payoff, they intend to impose β times the penalty on their opponent. An agent refreshes his strategy by referring to their own payoff and their neighbors' payoffs. In this study, we adopt the imitation-max (IM) update rule. This means that the focal player i imitates the strategy that has the maximum accumulated payoff of all strategies used by the player and their immediate neighbors, with strategy updating occurring simultaneously. By iterating the gaming and updating strategies, the global cooperation fraction, P_c, which is the ratio of agents adopting C to the total number of agents (N), is updated at each time step. This process is repeated until the global cooperation fraction attains equilibrium. For statistical reasons, we evaluate an ensemble average of 100 independent trials for each parameter set.

2.5.2 Results and Discussion

Figure 2.32 shows P_c, the averaged cooperation fraction, over 100 independent trials for different p values with fixed β (=1). $p = 0$ represents the default case, in which no jealous punishment is imposed. The general tendency in both panels A and B implies that cooperation is enhanced with increasing p. Since the cost–penalty ratio is kept at 1, greater damage is caused by a jealous sanction, which, although entailing

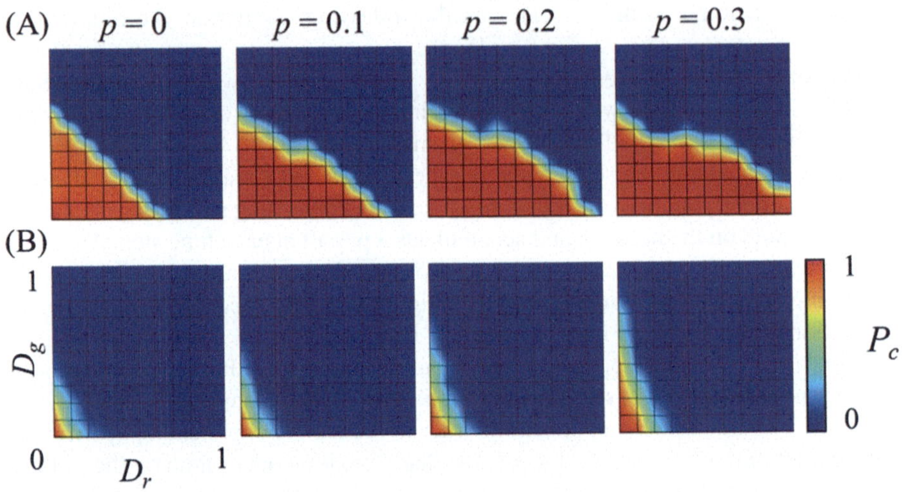

Fig. 2.32 Ensemble-averaged cooperation fraction, P_c, while varying the penalty ratio p with constant efficiency $\beta = 1$. The assumed spatial structures are (**a**) a lattice and (**b**) BA-SF

a larger cost, results in a positive effect that ultimately enhances cooperation. Note that in other cases that implement an update rule different from IM, such as a pairwise-fermi or random-underlying-network topology, no significant enhancement can be observed.

In the case of the lattice (panel A), more significant enhancement than that in the default case ($p = 0$) is observed when increasing the number of SH-type dilemmas (D_r) and decreasing the number of chicken-type dilemmas (D_g). In contrast to this, games with only Chicken-type dilemmas ($D_r = 0$) show only a slight improvement in terms of cooperation over the default case. This fact suggests that the enhancement caused by the jealous-punishment mechanism depends on the dilemma structure. Figure 2.33 indicates one of the representative time series showing who (A: the punisher) punishes whom (B: the victim) as A → B for each of the representative game structures with only Chicken-type or SH-type-dilemma games. In the former case, at the beginning of the episode, C to D is the most frequent occurrence. This is because a large D_g but low D_r gives a huge incentive to defectors exploiting cooperators. Additionally, cooperators punishing defectors must pay a punishing cost, giving them a significant disadvantage. In the end, cooperation perishes. In the latter case, however, D to D is dominant until the fifth time step, which puts defectors in a "dog-eat-dog" situation. Also, unlike in Chicken-type dilemma games, $0 = D_g < < D_r$ gives no merit to defectors exploiting cooperators, although cooperators must undergo an SH-type dilemma. These points work together to let cooperation survive and increase after the fourth time step, ultimately allowing cooperative equilibrium.

Next, let us focus on the case of BA-SF shown in Fig. 2.32b. Although the default setting shows a meagre cooperation level, cooperation gradually grows with increasing sanction strength. In a spatial game presuming a heterogeneous network like

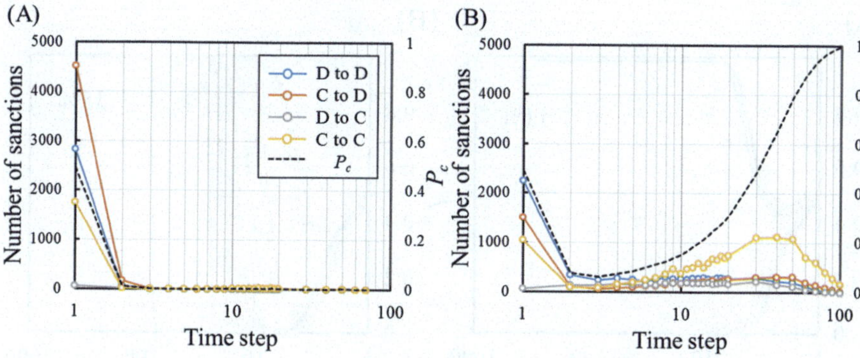

Fig. 2.33 Representative time series of the cooperation fraction and the types of sanctions in different two-dilemma structures when adopting a lattice. The parameter settings are $p = 0.2, \beta = 1$, and (**a**) a Chicken-type dilemma game ($D_g = 0.7$, $D_r = 0$), and (**b**) an SH-type dilemma game ($D_g = 0$, $D_r = 0.7$)

BA-SF, higher-degree (hub) agents play an important role in the dynamics. Generally, if defection is assigned to hub agents at an initial moment, D immediately spreads from those hub defectors to their subordinate cooperators. Hence, cooperation dies out at an early stage. In contrast, if cooperation supersedes defection and occupies a hub once, it will spread widely, replacing defection in a "cooperative avalanche." Figure 2.34 compares the default setting (panel A) with the jealous-sanction case (panel B), showing the 100 independent values of P_c. Without jealous sanction, cooperation has no chance of survival unless P_c is at a high level to begin with, requiring the major hubs to be cooperator-dominated at the start. Thus, the number of episodes reaching cooperative equilibrium is low. However, the case with jealous sanctions provides an opportunity for cooperation, even once severely oppressed by defection, to eventually recover over some time. Thus, panel B realizes quite a high frequency of cooperative equilibrium vis-à-vis panel A.

Based on each representative episode of high cooperation, successful cooperative equilibrium, and defector domination out of the 100 realizations shown in Figs. 2.34b and 2.35 shows the evolution of P_c and the types of sanctions observed among agents as in Fig. 2.33. Until the third time step, which shows an initial chaotic situation, there is no significant difference between the two panels. However, in the failed episode (panel A), cooperation dies out despite heavy sanctions given to defectors, while it does not perish in the successful case (panel B). The point of bifurcation between the two hinges on whether cooperators can form a C-cluster that can endure exploitation by the neighboring defectors. As observed in the successful case in panel B, if C-clusters can survive in the initial severe environment, this will be followed by a high rate of cooperators punishing defectors, allowing cooperation to gradually extend in two directions, into both the subordinate and superordinate defective regions in the network. When this gradual expansion of cooperation reaches a hub and switches it from defection to cooperation, an explosive expansion of cooperation is triggered, leading to a skyrocketing P_c, as observed in panel B.

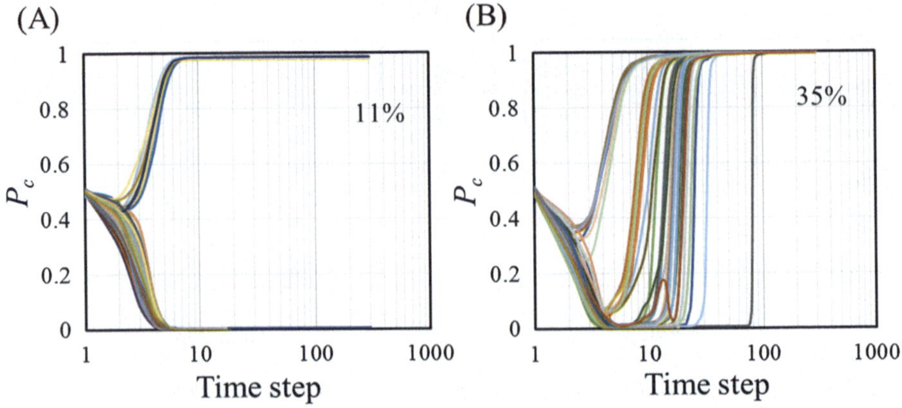

Fig. 2.34 Time series of the cooperation fraction for a 100-ensemble simulation with an underlying BA-SF network. The parameters are set as $D_g = 0.5$, $D_r = 0$. (**a**) Default (without sanctions; $p = 0$), (**b**) $p = 0.2$ and $\beta = 1.0$. The percentage values indicate the likelihood that an episode can attain a high cooperation level at equilibrium

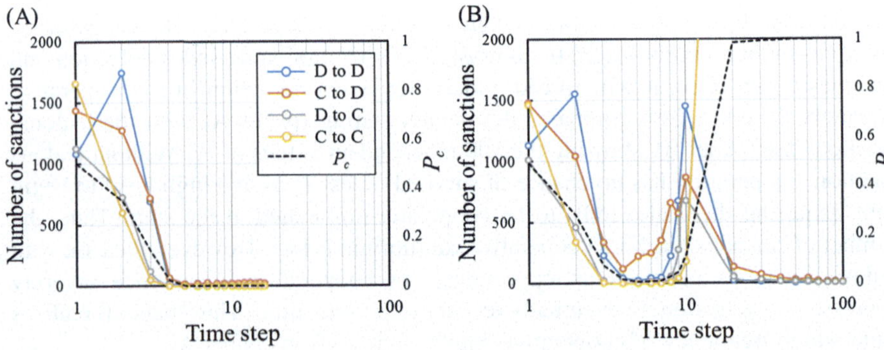

Fig. 2.35 Two representative time series showing the cooperation fraction and the prevalence of each type of sanction in the BA-SF case. The settings are $D_g = 0.5$, $D_r = 0$, $p = 0.2$, and $\beta = 1$. (**a**) A case in which cooperation goes extinct (a failed case), (**b**) a case in which cooperation prevails (a successful case)

The following discussion concerns how the efficiency of jealous sanctions affects the network reciprocity. Figure 2.36 shows the averaged P_c over the entire PD region (D_g, $D_r \in [0, 1]$), $P_{c|All \cdot PD}$ when varying the efficiency of the sanctions β and the penalty coefficient p in each network. As discussed in the previous paragraphs, more enhanced cooperation can be observed as the penalty coefficient increases, regardless of the underlying networks. In the lattice case, there is a tendency for higher efficiency to ensure more enhanced cooperation, although it is not as strong. Interestingly, however, in the BA-SF case, an inverse tendency is observed, whereby lower efficiency results in enhanced cooperation. This fact is justified below. Figure 2.37 shows what happens at the sixth time step of the episode shown in

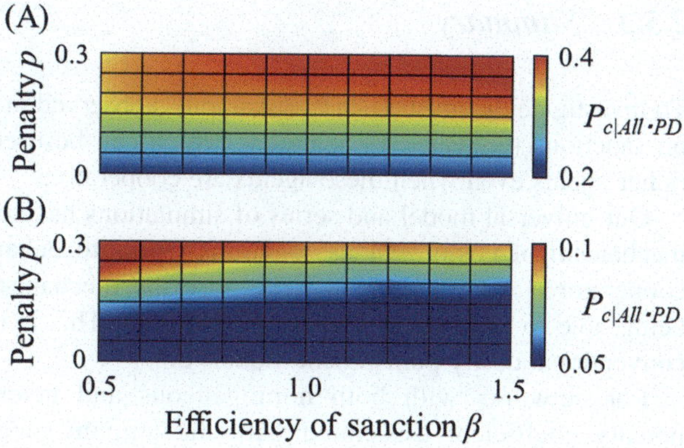

Fig. 2.36 Average cooperation fraction, P_c, over the whole PD area (D_g, $D_r \in [0, 1]$). We varied both the sanction efficiency β and the penalty ratio p. The spatial structures are (**a**) a 2D lattice and (**b**) BA-SF

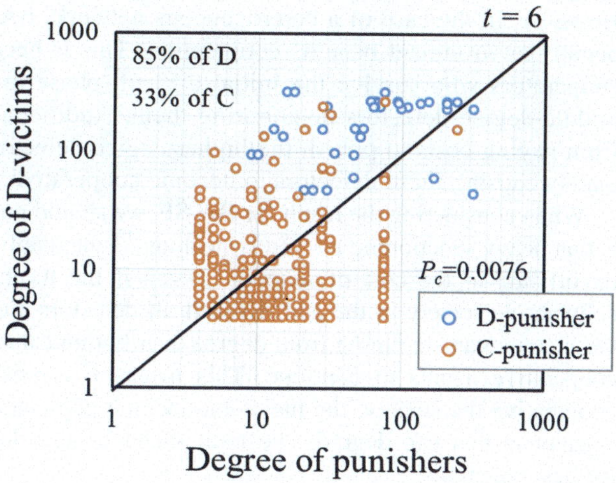

Fig. 2.37 Degree of punishers vs. degree of D-victims at the time step with the smallest number of cooperators ($t = 6$) in the episode shown in Fig. 2.4b. At this time step, P_c is observed to be 0.0076. As in Fig. 2.4, the assumed settings are $D_g = 0.5$, $D_r = 0$, $p = 0.2$, and $\beta = 1$. The percentage in the figure indicates how many agents of each strategy are sanctioning higher-degree defectors than themselves

Fig. 2.35b, wherein the jealous sanction successfully helps a cooperative equilibrium to emerge. Only sanctions of the C → D and D → D types are plotted. Obviously, cooperators tend to sanction defectors of a comparable degree level. By contrast, most instances of defectors sanctioning other defectors are over the +45-degree line, which implies that defectors tend to sanction other defectors of higher degree level or superordinate defectors in the network. This fact indicates that defectors of the middle-degree level, perhaps neighboring a C-cluster, tend to sanction neighboring cooperators of the same degree level and neighboring defectors of a higher degree level. Thus, those middle-degree defectors suffer penalties imposed by cooperators and the cost of punishing defectors at the same time. With worse efficiency, meaning the same sanction with a larger cost, the middle-degree defectors are more likely to copy cooperation from their neighbors. In a nutshell, this mechanism results from the heterogeneity of the underlying network.

2.5.3 *Summary*

To investigate the mystery of cooperation as observed in human society, we focused on sanctions motivated by jealousy, which can be used to reduce the payoffs of richer agents even when these agents are cooperative.

Our universal model and series of simulations have clarified several points. The mechanism of costly jealous sanctions works to enhance cooperation, even if a cooperator is punished. This is because the mechanism prevents defectors from being able to unilaterally exploit cooperators. This is basically the same as what conventional costly punishment models do.

For networks with both homogeneous and heterogeneous topologies, the penalty coefficient associated with the amount of damage is primarily what determines the evolution of cooperation. When a lattice is adopted as the underlying social network, cooperation is enhanced by an efficient cost–penalty ratio. However, in the case of a heterogeneous network, BA-SF, a less-efficient cost–penalty environment benefits cooperation. This is because heterogeneity allows cooperators who survive the initial difficult phase to reduce the payoff of the middle-degree defectors neighboring them. Additionally, these defectors suffer from paying costs to punish the higher-degree defectors surrounding them, ultimately causing these defectors to become cooperative as time evolves.

When considering the result for BA-SF, we should note one serious irony, which is that heavy sanctioning backed by jealousy[92] inevitably suppresses the total social payoff due to the cost of sanctions, even if the mechanism realizes an entirely cooperative society in the end. Even if all defectors are driven out of the society, the income gap stemming from degree distribution causes jealous sanctions among cooperative agents to continue. This means it is possible that, no matter how cooperative the society, the inevitable income gap among the population awakens a jealousy that can degrade the total social benefit due to wasting resources on fruitless sanctions.

This model provides a first step toward understanding the role of jealousy in the evolution of cooperation. An extended model that includes group jealousy will be a focus of our future work.

2.6 A Social-Dilemma Structure in Diffusible Public Goods

We discussed one of the most representative multiplayer games—the public-goods game (PGG)—in Sect. 2.2.2.2. Substantially, PGG is a multiplayer prisoner's dilemma game.

In this section, we will be concerned with an exciting application of PGG on the latest frontline of evolutionary game theory.

[92]Dercole et al. (2013).

PGGs can be used not only in economics but in several interdisciplinary fields. Let us briefly review the definition. An agent can adopt one of two strategies: cooperation (C) or defection (D). A cooperator (or C agent) offers to invest a certain amount of their funds (called a cost) into a common pool. The cumulative investment dedicated by the cooperators is multiplied by *r*, and this amplified benefit is distributed to all participants equally, irrespective of their strategies. Therefore, adopting the D strategy and taking a free-ride on the benefits must dominate, even though the system will collapse if all players do the same (the tragedy of commons[93]).

The dilemma situation associated with the PGG has been observed in the world of microbes.[94,95,96,97] It is well-known that the budding yeast species *Saccharomyces cerevisiae* helps others by producing invertase enzymes that hydrolyze sucrose to fructose and glucose. The monosaccharides (glucose and fructose) provided by hydrolysis are necessary resource for the survival of other types of yeast and are distributed via diffusion through the nutrient medium. In essence, it can be said that *S. cerevisiae* produces public goods while the other yeasts free ride. The relationship between the two types of microbes resembles the roles of the cooperator (donor) and the defector (recipient) in a social-dilemma game. Therefore, when they appropriately account for the diffusion of "public goods" or resources in a field, PGGs may fairly describe this particular social dilemma in the world of microbes. Several previous studies have reported that the dynamics of resource diffusion can be interpreted as a redistribution of public goods, which has led to the establishment of advanced PGG models including mathematical and cell-automaton models.[98,99,100,101,102,103,104,105] Note that this system is not entirely defined as the PD class; a snow-drift result has also been reported.[106,107,108]

Most notable is a study by Allen et al.,[101] who proposed a mathematical model to describe the social dilemma observed between the two types of microbes mentioned

[93]Hardin (1968).

[94]Velicer (2003).

[95]West et al. (2007).

[96]Julou et al. (2013).

[97]Koschwanez et al. (2013).

[98]Allison (2005).

[99]Driscoll and Pepper (2010).

[100]Momeni et al. (2013).

[101]Allen et al. (2013).

[102]Olejarz and Nowak (2014).

[103]Ermentrout and Edelstein (1993).

[104]Maire and Youk (2015).

[105]Celiker and Gore (2014).

[106]Gore et al. (2009).

[107]Doebeli and Hauert (2005).

[108]Greig and Travisano (2004).

above, *S. cerevisiae* and its followers (or free riders), as a generalized PGG in which the dilemma structure, amazingly, is consistent with the inequality associated with network reciprocity. Specifically, cooperation is favored only when the benefit-to-cost ratio is larger than the network degree, $b/c > k$, a condition that Nowak proved to be one of the five fundamental mechanisms of "social viscosity".[1] One novel point identified by Allen is that resource diffusion can be modeled by a transformed relation between the vertices connected by an edge, i.e., a network relation.

Despite its novelty, Allen et al.'s study seems less appropriate when applied to a general social dilemma, perhaps one universally applicable in the sphere of microbes, because in their model, a cooperator can maintain a constant resource-production rate, regardless of their inner metabolic level. It would be more appropriate if a cooperative-minded agent could not dedicate resources when exposed to a meagre environment, while being able to behave in a fully cooperative manner when faced with abundant primary resources. In other words, the Allen model potentially assumes a system with two unique resources, where the cooperator relies on a different primary resource than the resource that both cooperators and defectors consume to enhance their fitness.

The study presented below, addresses a single-resource system as opposed to a multi-resource one, and therefore defines only a single resource. Any agent can use this resource for survival, but a cooperator uses some part of the resource it imports from its environment to amplify it by means of an anabolic mechanism; it subsequently releases this product, which then diffuses throughout the field. We establish a new mathematical model and examine how rich phases can result. We investigated whether the population, which shares the resource through diffusion, goes extinct as in the tragedy of commons, or builds a sustainable system.

2.6.1 *Model and Methods*

We aimed to establish a general model for the evolutionary process of a species of microbes living on a plate of agar gel. Each microbe puts down its roots at a certain location in the 2-dimensional gel. The microbes therefore do not move on the gel, but rather the resource diffuses through the gel according to the archetypal 2D-diffusion equation. We can define the concentration of the resource at any point on the gel. The microbes can be divided into two categories. The first is the defective type (hereafter called defectors, D), which consumes a certain amount of the resource by absorbing it from the gel. The second is the cooperative type (hereafter called cooperators, C), which likewise consume the resource but simultaneously replenish the resource through a specific anabolic effect. The resource, produced in this way, is released into the gel. For simplicity, this study assumes that the only evolutionary process that takes place is in the turnover of single individuals, and that the evolutionary process of cell-division that results in the growth of the microbe colony into the surrounding vacant space does not occur.

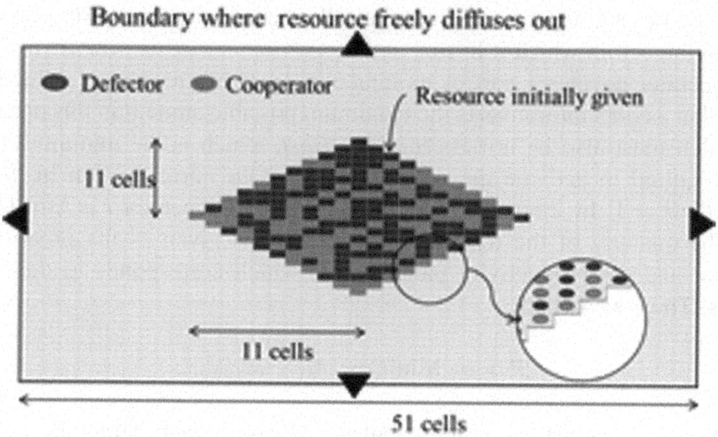

Fig. 2.38 Computational domain and cell positions. Initially, $N = 221$ agents are assigned to a diamond-shaped region at the center of the domain, where C (gray) and D (black) agents are randomly placed. The resource can flow out through the boundaries

2.6.2 *Mathematical Model of Diffusible Public Goods*

We assume a 2D gel space divided into N (=51 × 51; see Fig. 2.38) von Neumann cells. We initially place n_O (=221; see Fig. 2.38) microbes in the center of the gel region in a diamond formation, as shown in Fig. 2.38. Each microbe occupies a single cell of the gel. We express the address of a cell as an index i and assume that the concentration of the resource in that cell is given by Ψ_i. The strategy (C or D) of an agent located in cell i is denoted as $s_i \in \{+1,-1\}$. A cooperator (C: $s_i = 1$) produces and releases the resource P into its own cell. By contrast, a defector (D: $s_i = -1$) cannot produce the resource. Any arbitrary agent consumes a quantity of the resource, M_i, to maintain its fundamental metabolism. Denoting the equivalent overall diffusion coefficient by Λ, the governing equation can be written as follows:

$$\frac{d\Psi_i}{dt} = \Lambda\left(\sum_{j\in\{N_i\}} \Psi_j - \Psi_i\right) + \mathrm{Abs}[s_i] \cdot \left(\frac{s_i + 1}{2} \cdot P - M_i\right), \tag{2.43}$$

where $\{N_i\}$ indicates the set of four von Neumann cells neighboring cell i.

An agent, regardless of its strategy, imports the resource, which depends on the intake efficiency μ and the concentration difference between the internal resource concentration of agent i (Ψ_{body}) and Ψ_i. The resource intake must at least equal the basal metabolism, M_{min}. Thus, we can evaluate M_i as follows:

$$M_i = \mathrm{Max}\left[\mu\left(\Psi_i - \Psi_{body}\right), M_{\min}\right]. \tag{2.44}$$

An agent dies and is removed from the cell whenever it cannot import more than M_{min} from the cell in which it lives.

A cooperator devotes a part of its intake resources to resource production. We presume that a cooperator invests the maximum possible amount in this process, M_i-$M_{\min}$, which must also be no less than the *Cost*, which is the minimum resource amount required to activate the anabolic effect. Our normalization in this study assumes $Cost = 1$. In short, the amount invested by cooperator i is Min[*Cost*, M_i-$M_{\min}$]. The quantity of the resource produced, P, is given as the product of the investment and the productivity parameter b, which corresponds to the dilemma weakness. Thus, we obtain.

$$P = b \cdot \mathrm{Min}[Cost, M_i - M_{\min}]. \tag{2.45}$$

The fitness of agent i, π_i, which stipulates its evolutionary strength, depends on whether it is cooperative or defective as follows:

$$\pi_i = M_i - \frac{s_i + 1}{2} \cdot \mathrm{Min}[Cost, M_i - M_{\min}]. \tag{2.46}$$

2.6.3 Death–Birth Process

Stochastic-adaptation processes occur in microbe populations. In this study, we adopt a death–birth process to replicate an agent's evolutionary process.[109] At each discrete step, a randomly selected agent dies and is removed. Simultaneously, the fittest agent in the neighboring von Neumann cells (if any) is selected to produce a new offspring that refills the vacant cell. This offspring inherits its parent's strategy. If there are no neighbors, the cell remains vacant. Therefore, by performing this reproduction process based on a fitness evaluation, a dynamics occurs that results in a changing strategy profile in the system. If, at any point, only defectors remain, the evolutionary episode eventually ends in extinction, as the resource is not replenished in the domain.

2.6.4 Simulation Settings

In each episode, the initial resource is distributed to those cells to which an agent has been initially assigned, i.e., in the diamond formation mentioned above and shown in Fig. 2.38. The initial concentration is denoted as Ψ_{init}. According to the diffusion

[109]Ohtsuki et al. (2006).

equation in Eq. (2.43), the initial resource distribution spreads over time. Let us assume that each boundary of the domain adopts the free-boundary condition (in other words, that the concentration outside the domain is always maintained at zero), such that the resource can freely flow out. Therefore, if no cooperators remain in the domain, sooner or later the resource will run out and all of the agents will go extinct.

To provide a numerical solution for the diffusion equation, the so-called explicit method for time is applied. The fundamental parameters are set to $\Psi_{body} = 0$, $\Psi_{\text{init}} = 10.0$, $M_{\min} = 0.1$, and $Cost = 1$, and the control parameters are as follows: the diffusivity of resource Λ, the intake efficiency μ, and the dilemma weakness b. Moreover, in terms of the features expressed by public-good games with diffusion, mutation is not considered to avoid the complexity resulting from the increase of population size (i.e., a mutation rate of 0% is assumed).

In the following text, we discuss the equilibrium configuration that is achieved for a specific set of control-parameter values as a result of the competitive dynamics between C and D. Specifically, starting from a state with an equal number of Cs and Ds ($= n_0/2$) randomly placed in the diamond area in Fig. 2.38, we determine whether cooperators ultimately dominate, defectors dominate, or both species coexist. The defector-dominated state (all-defectors-state) is not sustainable and results in eventual extinction as no more of the resource is produced in the domain. Meanwhile, coexistence means that defectors can successfully free ride on the work of cooperators, and this is defined as a situation in which both C and D agents survive after an appropriately large number of time iterations (200,000 time steps).

By performing 300 independent trials, we obtain an ensemble-average value for the statistics. The result was obtained for various sets of the three parameters by varying the initial cooperation fraction.

2.6.5 *Preliminary Results*

Before reporting the competition-simulation results for the two species, to ensure the inherent sustainability of the population, we first conducted simulations to determine whether a population consisting of only cooperators could survive for various initial population densities, ρ. Specifically, starting from a certain ρ in which Cs are randomly assigned in the diamond in Fig. 2.38, we examined how many realizations of the 300 trials allowed the population to survive. The result, presented as the average frequency of extinction for each dilemma strength, is shown in Fig. 2.39. In both Fig. 2.39a1, b1, we can clearly see that the population can survive across most of the parameter range, except when μ is very close to 0. This implies that too small of a μ value does not even allow cooperators to survive due to the lack of sufficient nutrition. From the indicated values at the two cross-marks in the panels, we can see that Fig. 2.39a1, which assumed a stronger dilemma, has a slightly higher possibility of extinction. Examining the details in Fig. 2.39a2, b2, we see that the population is more likely to go extinct when assuming a smaller initial population density. Moreover, in Fig. 2.39a2, which

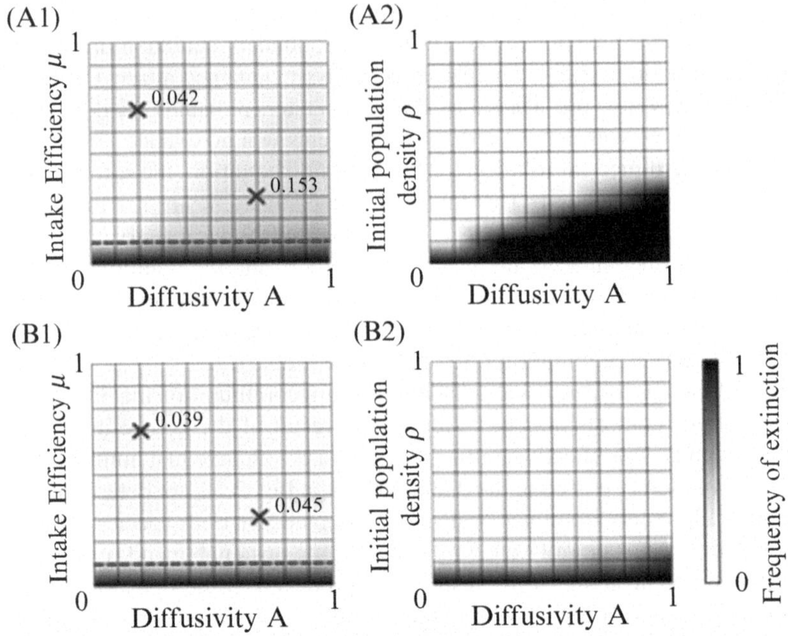

Fig. 2.39 The frequency of extinction for different dilemma strengths for 300 independent trials when the population is made up of only C agents. The average frequency of extinction in the Λ-μ plane is (**a1**) $b = 2$ and (**b1**) $b = 5$, where the arithmetic-average value for the initial population density ρ varying in the range [0, 1] is used. The numerical values of the marked crosses indicate actual frequencies. (**a2**) and (**b2**) show the frequencies of extinction in the Λ-ρ plane where $\mu = 0.1$ is assumed, and varying Λ is shown by the red dotted line in (**a1**) and (**b1**), where for (**a2**) $b = 2$ and for (**b2**) $b = 5$

assumes the stronger dilemma, depending on the diffusivity of the public goods, the extinction frequency increases even when a reasonably large initial population density is given. This implies that, even without exploiting agents, the cooperators cannot maintain their population by allowing the resource to diffuse into the surrounding space. This study does not account for the increasing number of agents in an episode from the viewpoint of exploring the inherent system dynamics brought about by public-goods diffusion; however, if we were to consider this, the possibility of the population surviving would be slightly higher.

2.6.6 *Results and Discussion*

In this section, we discuss the social-dilemma structure in public-goods games with diffusion, namely, whether or not the population survives as a result of competition between two species.

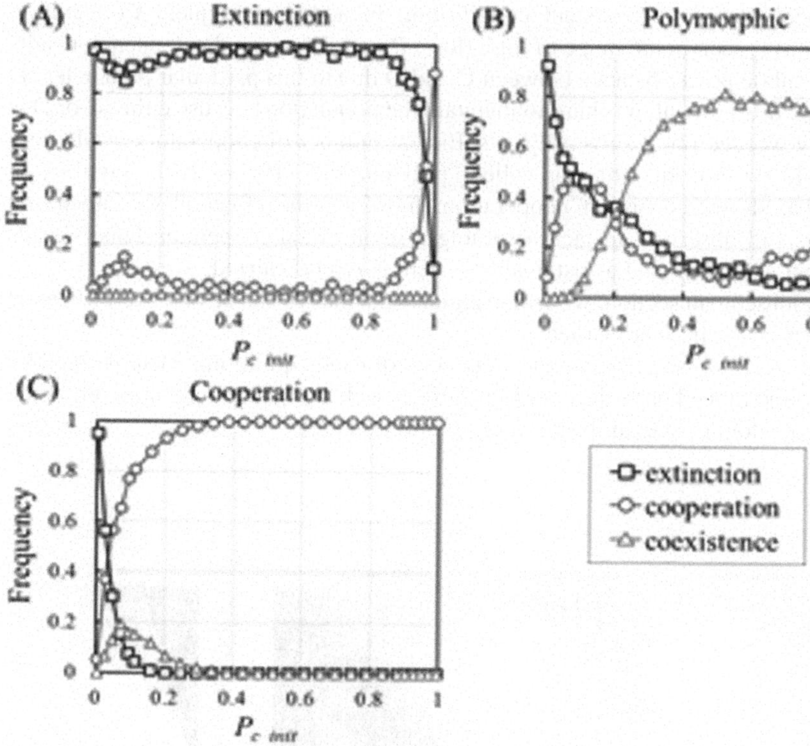

Fig. 2.40 Frequency of the three possible equilibria of the 300 independent trials for different initial cooperation fractions, P_{c_init}, with a constant initial population density $\rho = 1$: (**a**) Extinction ($b = 5$, $\Lambda = 0.1$, $\mu = 0.1$), (**b**) Polymorphic ($b = 2$, $\Lambda = 0.4$, $\mu = 0.8$), and (**c**) Cooperation ($b = 5$, $\Lambda = 0.5$, $\mu = 0.1$). Here, the square, circle, and triangular symbols show the frequency of each equilibrium: extinction, cooperation, and coexistence (Polymorphic), respectively

Under a particular parameter set for Λ, μ, and b, we can compile statistics, expressed by the frequency of extinction (either true extinction or only defectors surviving), polymorphic (co-existing cooperators and defectors), or cooperation (cooperators dominating the system) results over 300 independent simulation trials while varying the initial cooperation fraction. In a typical illustration, Fig. 2.40 shows the equilibrium-frequency distribution. In observing Fig. 2.40a, we can see that extinction is prevalent and does not depend upon the initial cooperation fraction (hereafter P_{c_init}). However, when most agents are C (P_{c_init} is close to 1), this is not true; the frequency of cooperation exceeds that of extinction, which is inevitable because the stochastic-adaptation death–birth process is adopted. We classify this case as Extinction. In Fig. 2.40b, even though we can confirm that the system reaches various equilibria according to P_{c_init}, intermediate P_{c_init} values show a high probability of a coexistent equilibrium. In addition, frequent Cooperation (Extinction) is seen in the range of high (low) P_{c_init}. This complex tendency results from the subtle power balance between C and D due to this particular parameter set and P_{c_init}. If C is small, it is hard to maintain the population because either C or D is inevitably weeded out. Conversely, a sufficient number of cooperators results in a sustainable system; this phase is called "polymorphic". Figure 2.40c, contrary to Fig. 2.40a, shows consistent cooperation; however, the effect of the stochastic process is still observed. In fact, assuming a small initial cooperator density leads to frequent extinction and a sustainable system is rarely realized.

In the subsequent section, we will examine what phases can be observed when the parameters Λ, μ, and b are varied.

In Fig. 2.41, we see the average frequency of each equilibrium state defined by the above-mentioned procedure for Fig. 2.40, as well as a phase diagram represented by the most-dominant equilibrium (e.g., see Ref. [112]).

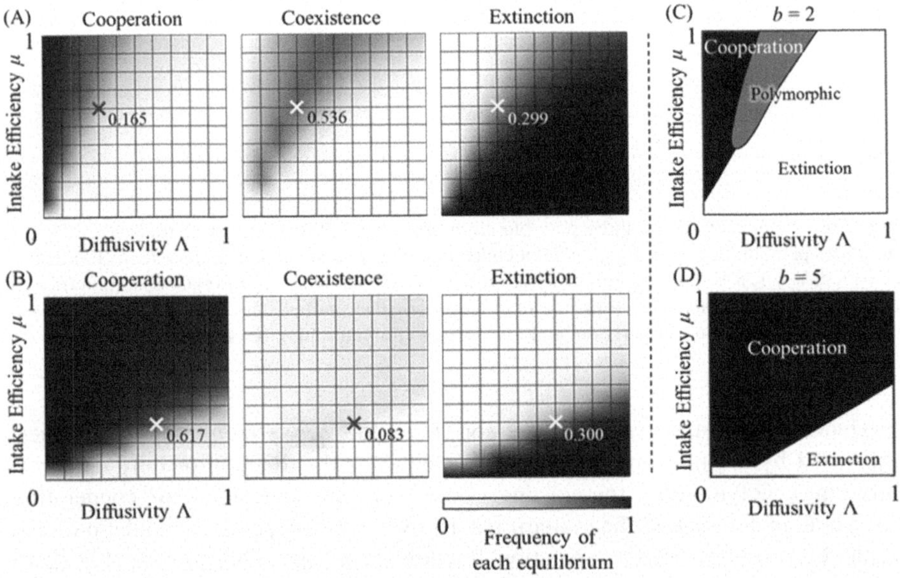

Fig. 2.41 The frequency of each equilibrium and the phase diagrams of the dynamics drawn in the Λ-μ plane for different dilemma strengths. The fixed parameter for (**a**) and (**C**) is $b = 2$; for (**b**) and (**d**), it is $b = 5$. Panels (**a**) and (**b**) show the equilibrium frequency for cooperation, coexistence, and extinction with varying initial cooperation fractions with $\rho = 1$. The numerical values in the panels indicate the actual frequency at the settings indicated by the cross-marks. Panels (**c**) and (**d**) show phase diagrams of the dynamics based on Panels (**a**) and (**b**), where only the frequency of the most frequent equilibrium is represented and coexistence is labeled as Polymorphic

2.6.7 Strong Dilemma

As shown in Fig. 2.41c, the phase transition C-dominated -> Polymorphic -> D-dominated takes place in a sequential manner with increasing diffusivity and is more obvious than in the case of a weak dilemma, as in Fig. 2.41d and discussed below. If the resource spreads slowly throughout the domain with a smaller diffusivity, a cooperator will have access to a relatively large quantity of the resource, whereas a defector who relies on the resource flowing from the surrounding cooperators will have a more difficult time. This means that the fitness of the cooperators naturally exceeds that of the defectors, and that the cooperators' offspring will ultimately dominate the population. Consequently, C-domination emerges. If the resource diffuses reasonably quickly throughout the domain, defectors can more easily obtain the resources provided by the cooperators, allowing defectors to free-ride on the work of cooperators, leading them to produce offspring. However, the continued existence of cooperators is requisite for the survival of the defectors. Because both types of agents can coexist without extinction, a Polymorphic phase occurs. That is, the cooperators form a cluster and produce abundant resources with mutual cooperation, while the defectors obtain fewer resources at a distance. Accordingly, once the parameter balance collapses, the equilibrium state is pulled into either the cooperation or extinction phases. If the resource diffuses too quickly, the defectors become completely dominant, and the cooperators are replaced. This is because it is harder for a cooperator to keep the resource it produces for itself, and it must pay a production cost. In this case, the entire population eventually goes extinct because the complete absence of cooperators is not sustainable. In other words, "a society of free-riders with no tax-payers goes to ruin."

2.6.8 Weak Dilemma

In Fig. 2.41d, by improving the productivity and therefore relaxing the dilemma strength, a tendency for Cs to prosper over Ds is observed. In general, both low diffusivity and high intake rates work in favor of survival for cooperators, as was observed in the strong-dilemma case. This means that the polymorphic phase observed between the cooperation and extinction cases in Fig. 2.41c disappears because the relaxed dilemma favors cooperation; therefore, no free riders survive in the system as long as a smaller diffusivity is assumed. This is why the coexistence fraction is very small in Fig. 2.41b. Along with the relaxed dilemma from $b = 2$ to $b = 5$, the coexistent results decrease gradually. If one increases the diffusivity, despite the weaker dilemma, the dynamics suddenly shift from cooperation to extinction. As the social dilemma becomes weaker than $b = 5$, the cooperation phase is expected to enlarge and the extinction phase is expected to diminish. In general, the system favors one of two extreme equilibria, either cooperation or extinction, and is much less likely to adopt a "cloudy" equilibrium, such as the polymorphic phase. Therefore, realizing a sustainable system implies an all-cooperators state and does not allow defectors to survive. This makes us draw an interesting implication from social point of view, which is that 'fluid-like mobility of public good' such as a resource diffusing like what this study presumed makes difficult free-rider survive with exploitation from cooperative agents.

2.6.9 Summary

To reveal the underlying dilemma structure in a PGG in which the "public goods" diffuse throughout the system, we developed a new PGG model to reproduce the struggle between cooperators, who replenish a resource, and defectors, who free ride on the productivity of the cooperators.

To ensure sustainability at a certain level, preliminary simulations assuming that a population composed entirely of cooperators were conducted. We confirmed that the population could be sustained except when it had a small initial density, and variations were observed due to the effect of diffusivity.

Our numerical simulations resulted in phase diagrams where the diffusivity, the intake rate of the resource, and the dilemma strength were all varied. Despite the payoff function based on the PGG, the effect of the spatial structure (the diffusion and concentration distribution) resulted in very different outcomes from those estimated from the payoff function, i.e., the cooperation, polymorphic, and extinction phases. We found that a stronger social dilemma requires a smaller diffusivity to maintain the population because the produced resources are very small. That is, fast diffusion encourages the survival of defectors and a sustainable population cannot be achieved. An intermediate diffusivity establishes a symbiotic relationship, namely, clustering cooperators produce large quantities of resources and a few defectors get a free ride. A weaker social dilemma not only favors cooperation but also discourages "cloudy" equilibria, such as the Polymorphic phase, in favor of "clear" equilibria, either C-dominated or D-dominated.

As mentioned in the text, the present model assumes that the only evolutionary process taking place is in the turnover of individual agents and disregards another important evolutionary process, cell-division, which results in the growth of the microbe colony into the surrounding vacant space. In addition, it is expected that the number of individuals will experience fluctuations, such as those seen in Lotka–Volterra systems. At the beginning of each episode in such a system, defectors overtake cooperators and population size diminishes due to die-outs. In the next stage, cooperators in clusters generate new individuals with ample resources and the population size grows. However, defectors again exploit cooperators and increase in number, and the decrease in the population size is repeated. Thus, a certain sustainability, as seen in biological systems, can be produced. This type of system will be examined further in a subsequent report.

2.7 Three-Strategy Game: Cooperator, Defector, and Costly Cooperative Loner

EGT, in its typical framework, deals with 2-player–2-strategy games (i.e., 2×2 games). In such games, a player decides his strategy—cooperation (C), or defection (D)—in one-shot situations. The PD game, has been preferentially adopted as a metaphor for studying real social dilemmas, where cooperation is shown to be an unfavorable scenario unless reciprocity mechanisms are considered. Many cooperation-enhancing mechanisms have been studied. In this section, we discuss a three-strategy game. If adding a certain third strategy helps cooperation to survive, such a framework enables a reciprocity mechanism to be implemented. One example is the addition of the loner-strategic type (L) to represent players who do not participate in the game and are satisfied sharing small payoffs (σ) with their opponents.[110] These studies concluded that the evolutionary picture of PD games can be significantly altered by the addition of L.

It has been analytically proven that, as long as the dilemma strength is small, a three-strategy game in which C, D, and L coexist, can oppress exploiters, thereby avoiding the situation wherein all players mutually defect.[111] When an L declines from participating in the game, it hampers the manner in which D exploits C. Thus, Ls protect Cs by expelling Ds from the population. By contrast, Cs are dominated by Ds and dominate Ls. This triangular relation results in cyclic dynamics similar to those seen in the game rock–paper–scissors, which is one of the archetypical three-strategy games in which some stochastic perturbation exists.[112]

However, concern arises regarding the validity of L as a third strategy in real contexts. Although L indirectly hampers D's prosperity, it seems neutral in either helping C or harming D. Alternatively, the strategy known as costly punishment (P) can be invoked as a means of inhibiting the growth of defection.[113] The costly punisher acts as C in principle, cooperating with either C or P. However, P punishes D by imposing a penalty (γ) at a cost (δ). P actively cleans up defectors, differing from the loner game. In contrast, the loner game may encounter an analogous situation by establishing cyclic dynamics under some level of perturbation because introducing P as a third strategy can overwhelm D through parameters γ and δ.

In this section, the study introduces a new three-strategy game that combines the loner and costly punishment games, by introducing a costly cooperative loner (CL). As a third strategy, CL seems more realistic and plausible than either L or P because CL basically cooperates when interacting with another cooperator and not with a

---

[110]Following literatures constitute representative works. Michor and Nowak (2002), Jeong et al. 2014, Schoenmakers et al. (2014), Garcia and Traulsen (2012).

[111]Hauert et al. (2002).

[112]Szolnoki et al. (2014).

[113]Some representative works are as follows. Fehr and Gächter (2001), Henrich et al. (2001), Rand et al. (2009), Sekiguchi and Nakamaru (2009), Rand et al. (2010), Szolnoki et al. (2011b), Shinao and Nakamaru (2013).

defector. Hence, CL is a conditional cooperator similar to TFT.[114] It is, however, inevitable that CL will pay an additional cost (a). In other words, CL strategy can be compared with the game-exit option,[115] but at a certain cost.

2.7.1 Model Framework

In a 2×2 PD game, a player receives a reward (R) for each mutual cooperation (C) and a punishment (P) for each mutual defection (D). If one player chooses C and the other chooses D, the latter obtains a temptation payoff (T) and the former is labeled sucker (S). Without loss of generality, we can define a 2×2 game space by assuming $R = 1$ and $P = 0$ as follows:

$$\begin{pmatrix} R & S \\ T & P \end{pmatrix} = \begin{pmatrix} 1 & -D_r \\ 1 + D_g & 0 \end{pmatrix},$$

where $D_g = T - R$ and $D_r = P - S$ imply chicken- and SH-type dilemmas, respectively ([3], [8]). We limit the PD game class by assuming $0 \leq D_g \leq 1$ and $0 \leq D_r \leq 1$. One defines the payoff matrix, $\mathbf{M}$, as described in Table 2.5 (a). According to the original meaning of "cost"(a), we limit $0 \leq a \leq 1$.

Let us presume an infinite and well-mixed population whose strategy fraction (f_C, f_{CL}, f_D) is .

Assuming that the corresponding pure strategies are $\mathbf{x}_1 \equiv \begin{pmatrix} 1 \\ 0 \\ 0 \end{pmatrix}$, $\mathbf{x}_1 \equiv \begin{pmatrix} 0 \\ 1 \\ 0 \end{pmatrix}$, and $\mathbf{x}_3 \equiv \begin{pmatrix} 0 \\ 0 \\ 1 \end{pmatrix}$, the replicator equation is derived as follows:

$$\frac{\dot{x}_i}{x_i} = \left({}^T\mathbf{xi} \cdot \mathbf{Mx} - {}^T\mathbf{x} \cdot \mathbf{Mx}\right), \tag{2.47}$$

where the superscript T indicates the transpose operator and subscript $i \in \{1, 2, 3\}$. Note the following constraint:

$$x_1 + x_2 + x_3 = 1. \tag{2.48}$$

At equilibrium, Eq. (2.47) reads as follows:

[114] Axelrod (1984).

[115] Schuessler (1989).

$$0 = x_i \cdot \left({}^T\mathbf{xi} \cdot \mathbf{Mx} - {}^T\mathbf{x} \cdot \mathbf{Mx}\right). \tag{2.49}$$

By substituting Eq. (2.48) into Eq. (2.49) to eliminate x_3, a simultaneous cubic equation with two unknown variables (x_1 and x_2) is obtained. Although such an equation generally has $3^2 = 9$ solutions as maximum, in this case, six equilibria are obtained. Three trivial equilibria are $(1 \quad 0 \quad 0)$ (Equilibrium #1), $(0 \quad 1 \quad 0)$ (Equilibrium #2), and $(0 \quad 0 \quad 1)$ (Equilibrium #3). Three other, non-trivial equilibria can be derived analytically as follows:

$$\text{Equilibrium \#4}: \left(\frac{D_r}{-D_g + D_r} \quad 0 \quad \frac{-D_g}{-D_g + D_r}\right),$$

$$\text{Equilibrium \#5}: \left(\frac{-a + D_r - aD_r}{(1 + D_g)D_r} \quad \frac{-aD_g + aD_r + D_gD_r}{(1 + D_g)D_r} \quad \frac{a}{D_r}\right),$$

$$\text{Equilibrium \#6}: (0 \quad a \quad 1 - a).$$

Equilibrium #4 always appears on the border line connecting the all-cooperators state, $(1 \quad 0 \quad 0)$, and the all-defectors state, $(0 \quad 0 \quad 1)$. Equilibrium #4 becomes $f_C < 0$ or $f_D < 0$ when obeying the PD game conditions ($0 \le D_g \le 1$ and).

More precisely, if $D_g > D_r$ (chicken-type dilemma stronger than SH-type dilemma), Equilibrium #4 brings $f_C < 0$, in contrast to the situation when $D_g < D_r$ (SH-type dilemma stronger than chicken-type dilemma), where the conditions for Equilibrium #4 are met when $f_D < 0$.

Equilibrium #6 must always appear somewhere on the border line connecting the all-costly CL state, $(0 \quad 1 \quad 0)$, and the all-defectors state, $(0 \quad 0 \quad 1)$. In a costlier situation($a \cong 1$), when CL encounters difficulties in surviving due to the large cost to exit the game, Equilibrium #6 closes to $(0 \quad 1 \quad 0)$.

Compared with the abovementioned equilibria, Equilibrium #5 is specific (the so-called internal equilibrium under certain situations) and appears in the meaningful strategy space only if within the triangle defined by $(1 \quad 0 \quad 0)$, $(0 \quad 1 \quad 0)$, and $(0 \quad 0 \quad 1)$.

Tables 2.6a, 2.6b, and 2.6c Parameters used in the definition of the payoff matrix, **M**. Assume a PD game limited to the following conditions: $0 \le \mathrm{D_g} \le 1$ and $0 \le \mathrm{D_r} \le 1$.

Table 2.6a Costly cooperative loner (CL) as a third strategy (as proposed herein)

Fraction	Strategy	C	CL	D
x_1	C	1	1	$-D_r$
x_2	CL	$1 - a$	$1 - a$	$-a$
x_3	D	$1 + D_g$	0	0

Table 2.6b Simple loner (L) as a third strategy; assuming $\sigma < 1$

Strategy	C	L	D
C	1	σ	$-D_r$
L	σ	σ	σ
D	$1 + D_g$	σ	0

Table 2.6c Costly punisher (P) as a third strategy; assuming $\delta < \gamma$

Strategy	C	P	D
C	1	1	$-D_r$
P	1	1	$-D_r - \delta$
D	$1 + D_g$	$1 + D_g - \gamma$	0

2.7.2 Trajectory of the Evolutionary Path

We numerically analyze the replicator Eq. (2.47) by varying the game's structural parameters, D_g, D_r, and a.

Figure 2.42a shows the vector maps and the color contours of the social average payoff in the strategy space ($f_C - f_{CL} - f_D$) at five representative settings of D_g and D_r ((* − 1), $D_g = 0.9$ and $D_r = 0.1$; (* − 2), $D_g = 0.1$ and $D_r = 0.1$; (* − 3), $D_g = 0.5$ and $D_r = 0.5$; (* − 4), $D_g = 0.9$ & $D_r = 0.9$; and (* − 5), $D_g = 0.1$ and $D_r = 0.9$) with varying five conditions of a (Figs.2.42b, 2.42c, 2.42d and 2.42e). Each vector drawn from $\boldsymbol{f}_{point} = \begin{pmatrix} f_C \\ f_{CL} \\ f_D \end{pmatrix}$ in the vector map is obtained by

$$\begin{bmatrix} (\mathbf{1} \ \ \mathbf{0} \ \ \mathbf{0}) \cdot \mathbf{M} \begin{pmatrix} f_C \\ f_{CL} \\ f_D \end{pmatrix} - (f_C \ \ f_{CL} \ \ f_D) \cdot \mathbf{M} \begin{pmatrix} f_C \\ f_{CL} \\ f_D \end{pmatrix} \\ (\mathbf{0} \ \ \mathbf{1} \ \ \mathbf{0}) \cdot \mathbf{M} \begin{pmatrix} f_C \\ f_{CL} \\ f_D \end{pmatrix} - (f_C \ \ f_{CL} \ \ f_D) \cdot \mathbf{M} \begin{pmatrix} f_C \\ f_{CL} \\ f_D \end{pmatrix} \\ (\mathbf{0} \ \ \mathbf{0} \ \ \mathbf{1}) \cdot \mathbf{M} \begin{pmatrix} f_C \\ f_{CL} \\ f_D \end{pmatrix} - (f_C \ \ f_{CL} \ \ f_D) \cdot \mathbf{M} \begin{pmatrix} f_C \\ f_{CL} \\ f_D \end{pmatrix} \end{bmatrix}.$$

The social-average payoff decreases with increasing a, which is quite conceivable and inherently results from the game definition (larger a imposes a more severe situation for CL when competing with D). Although the vector map may indicate that cyclic fluctuation exists amid the three strategies, we should note that any evolutionary path would consequently be absorbed by an all-defectors state, as shown in Fig. 2.43a. Equilibria #6 and #5, for some set of parameters, play an important role in the latter observation. With increase of a, an evolutionary path can more become easily trapped by the state with no cooperative strategies. In this sense, the observed result is plausible from a qualitative point of view, although counter-intuitive (Fig. 2.43b).

Figure 2.43a shows the trajectory unfolded from several initial boundary conditions. Panels (A) and (B) display the results for $a = 0.1$ and 0.5, respectively. Each

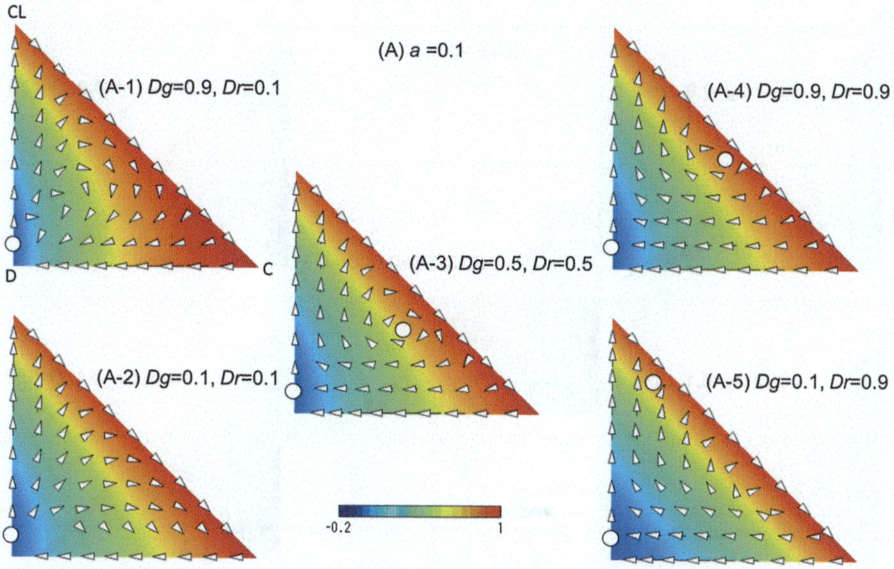

Fig. 2.42a Vector maps for the time-evolving strategy fractions from$\boldsymbol{f}_{\boldsymbol{point}} = (f_C \quad f_{CL} \quad f_D)$ and the color contour of the social-average payoff assuming initial conditions$\boldsymbol{f}_{\boldsymbol{point}} = (f_C \quad f_{CL} \quad f_D)$ and $a = 0.1$.Equilibria 5 and #6 are shown by the two open circles

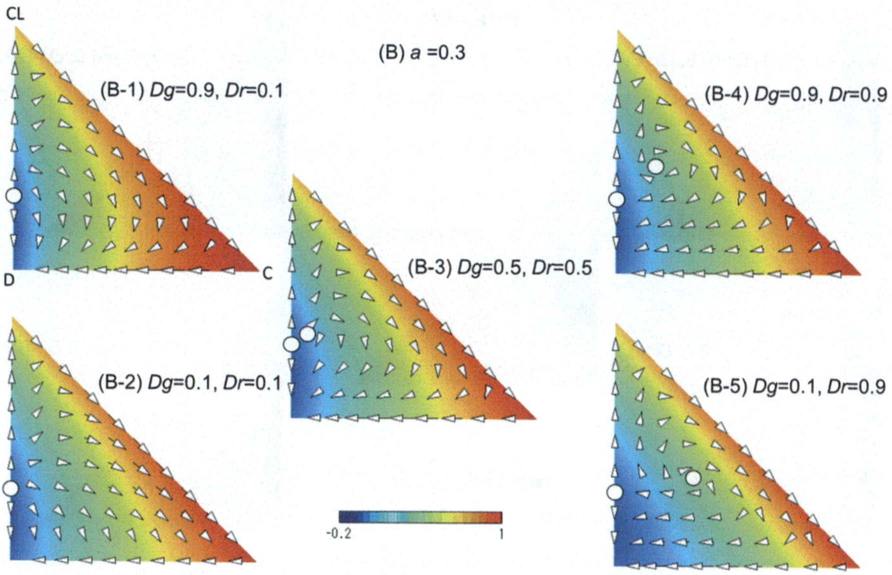

Fig. 2.42b Assuming $a = 0.3$, the results are presented in the same format as in Fig. 2.42a

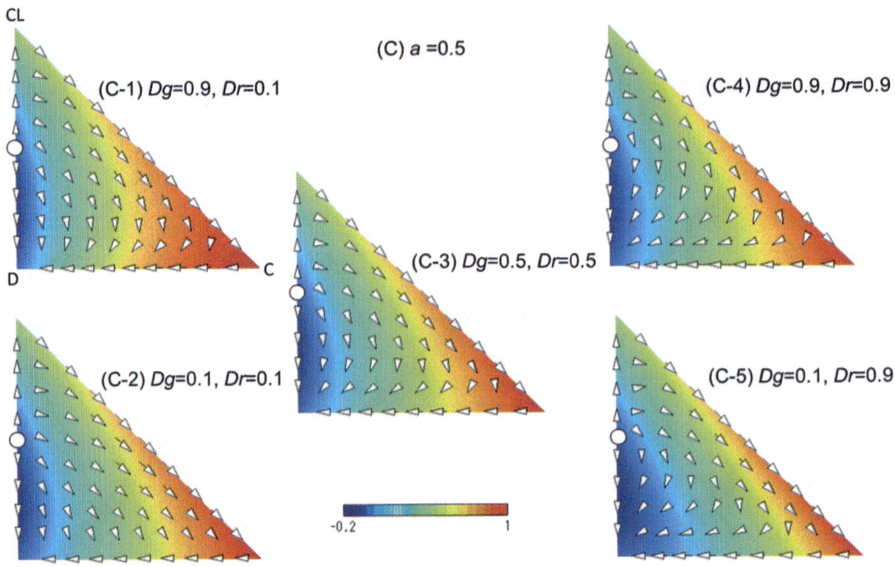

Fig. 2.42c Assuming $a = 0.5$, the results are presented in the same format as in Fig. 2.42a

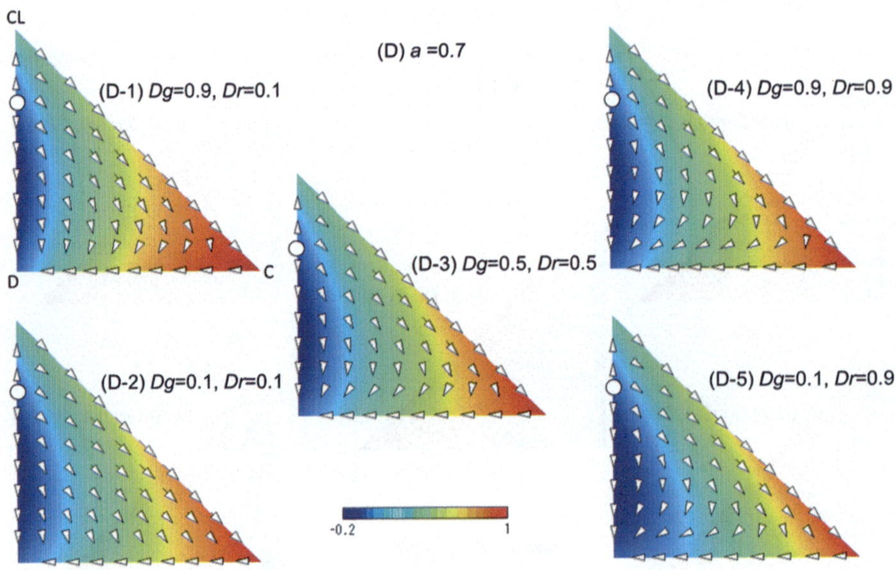

Fig. 2.42d Assuming $a = 0.5$, the results are presented in the same format as in Fig. 2.42a

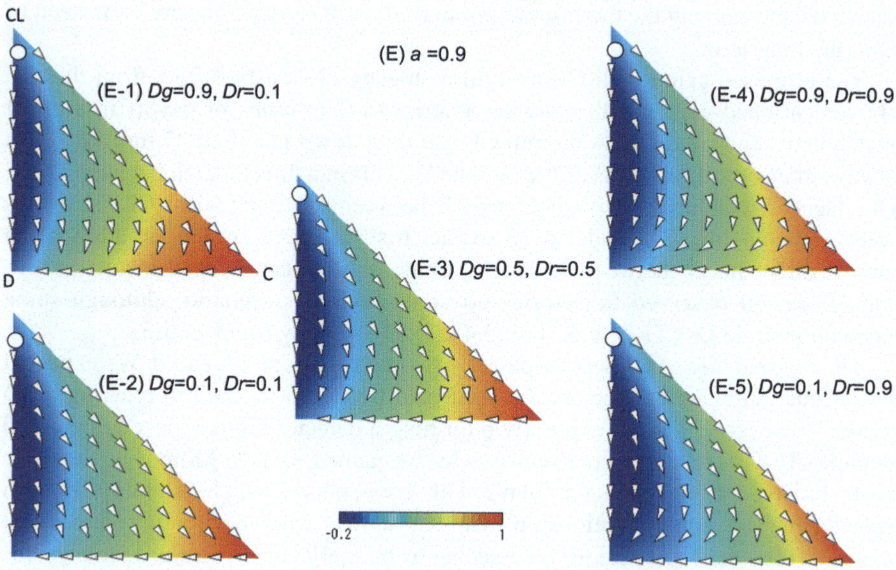

Fig. 2.42e Assuming $a = 0.9$, the results are presented in the same format as in Fig. 2.42a

panel contains the five settings of D_g and D_r ((* − 1)–(* − 5)) that are consistent with Fig. 2.42a. It confirms the ultimate all-defectors state in all cases.

Notably, all trajectories are, sooner or later, absorbed by the all-defectors state $(0 \quad 0 \quad 1)$. Even when setting small a, which allows CL to exit the game easily, CL is not able to survive and thus goes extinct along with C. Consequently, in every case, defection becomes dominant. This is always true even if low dilemma strengths are imposed. This observation means that the present 3-strategy game with costly CL does not change the evolutionary picture observed in a canonical 2×2 PD game, in which any evolutionary path is always absorbed by an all-defectors state as long as an infinite and well-mixed population is assumed. Regarding whether the introduction of a third strategy besides C and D can relax a PD situation, the present game (definition of parameters given in Table 2.6a) was revealed to be less successful than either the 3-strategy game with L (given by Table 2.6b; hereafter referred as the "loner game") and or that with Punisher (given by Table 2.6c, hereafter referred as the "punishment game").

A qualitative reasoning is proposed to justify our observations. First, L can benefit either the cooperator or the defector depending on σ in the loner game. In a nutshell, the loner game allows L to profit while others fight. By contrast, the punishment game relies on a different mechanism. Depending on the set of costs and fines, there can be a subtle balance amid the three strategies, where D exploits C, but P punishes D and C equals P. Although it is not entirely the same as the so-called three-way standoff mechanism typically observed in a rock–paper–scissors game, such a balance among the three strategies may allow P and/or C to survive instead of D in the long term.

The proposed game with CL as a third strategy obviously differs from the two abovementioned games. CL is always inferior to C in terms of payoff in a direct match between C and CL. Although CL can drag down D to help C survive, CL is always inferior to D in terms of the obtained payoff in a direct match between D and CL. Hence, with respect to direct match between two strategies, CL is always weaker. Inevitably, CL tends to go extinct first, followed by the situation of an almost direct match between C and D. Thus, only defectors exist in the end. In Fig. 2.43a, all observed trajectories go along with this scenario, although some fluctuation amid D–CL–C (e.g., Fig. 2.43aA-3, A-4) may occur in time.

The original idea of CL was inspired by two strategic types: L and P. We expected Ls to function against defectors at a cost paid by CL to succeed in relaxing a PD game. However, CL is not explicitly punishing a defector but hampers defectors to exploit CL passively (the core reason why we named it "L"). Moreover, there are costs involved even when CL plays with cooperators, which inhibits CL from establishing an equal relationship with C, unlike what occurs in punishment games. Thus, the CL strategic type seems to be ineffective as a third strategy for mitigating a PD situation.

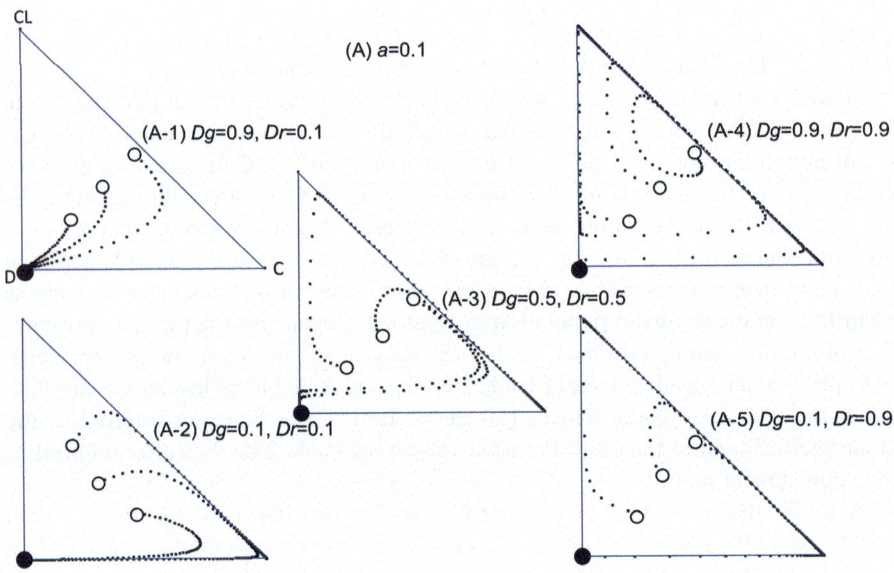

Fig. 2.43a Trajectories from different initial conditions. The open circles represent respective initial conditions

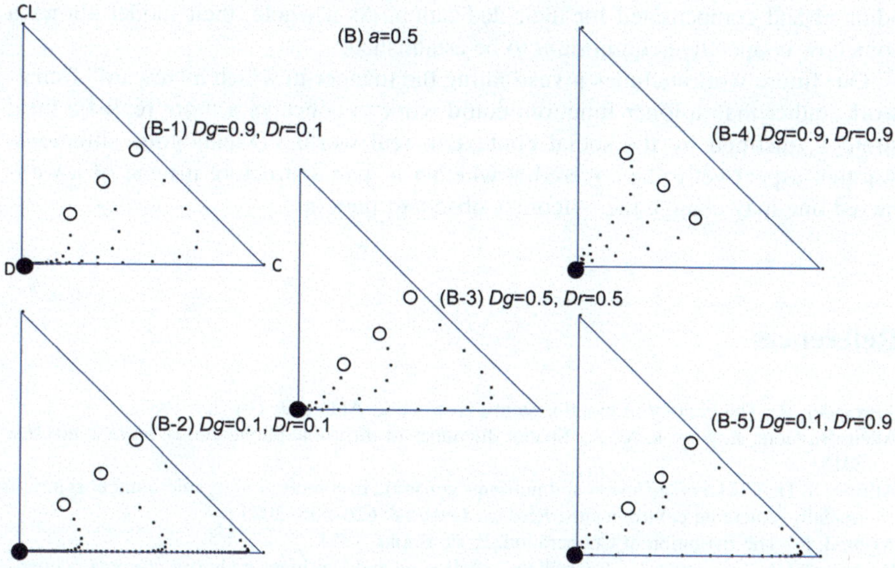

Fig. 2.43b Trajectories from different initial conditions. The open circles represent respective initial conditions

2.7.3 Discussion

In this section, we defined a new third strategic type: the costly CL, inspired by the L's function (game-exit option) and the punishment framework. Our analysis revealed that the introduction of CL does not relax the PD situation predominantly because (i) CL is less powerful in foiling D than the traditional costly punisher strategic type and (ii) CL is not entirely independent from the other two strategies, as in the case of a conventional loner strategy, thus enabling only an imperfect game-exit option.

Recently, Wang et al. reported a new framework of multiplayer and 3-strategy games based on the public-goods game, where C, D, and P are assumed as strategies; however, each of the three strategic types are given a different "reward". Depending on the controlling parameter of this "reward," three different evolutionary outcomes can emerge: (1) P-absorbing; (2) bi-stable equilibrium of P-absorbing or D-absorbing; and (3) bi-stable equilibrium of D-absorbing or a polymorphic sink of P and C.[116] Because of the introduction of such reward framework, P is given an advantageous status over C (as well as over D). This could be justified by our social consensus that cooperators who intentionally practice costly punishment should be

[116]Wang et al. (2018).

admired and compensated for their dedication. As a whole, their model allows a somehow cooperative equilibrium to be established.

Our future work includes investigating the manner in which a "reward" framework, rather than a loner function, could work to conceive a more realistic third strategy, justified by the social context in real world PD-analogous situations. Another aspect yet to be covered is whether a spatial structure instead of a well-mixed one may change the outcomes observed therein.

References

Alexander, R.; The biology of moral systems, New York: Aldine De Gruyter, 1987.

Allen, B., Gore, J., Nowak, M. A.; Spatial dilemmas of diffusible public goods, *eLife* **2**, e01169, 2013.

Allison, S. D.; Cheaters, diffusion and nutrients constrain decomposition by microbial enzymes in spatially structured environments, *Ecology Letters* **8**, 626–635, 2005.

Axelrod, R.; The Evolution of Cooperation. Basic Books, 1984.

Baldassarri, D., Grossman, G.; Centralized sanctioning and legitimate authority promote cooperation in humans, *Proceedings of the National Academy of Sciences of the United States of America* **108**, 11023–11927, 2011.

Barabasi, A.L., Albert, R.; Emergence of scaling in random networks, *Science* **286**, 509–512, 1999.

Benzi, R., Sutera, A., Vulpiani, A.; Wang, Z.; The mechanism of stochastic resonance, *Journal of Physics A: Mathematical and General* **14 (11)**, L453-L457, 1981.

Boyd, R., Gintis, H., Bowles, S.; Coordinated punishment of defectors sustains cooperation and can proliferate when rare, *Science* **328**, 617–620, 2010.

Brandt, H., Sigmund, K.; Indirect reciprocity, image scoring, and moral hazard, *Proceedings of the National Academy of Sciences of the United States of America* **102**, 2666–2670, 2005.

Brandt, H., Hauert, C., Sigmund, K.; Punishment and reputation in spatial public goods games, *Proceedings of the Royal Society of London B* **270**, 1099–1104, 2003.

Brandt, H., Hauert, C., Sigmund, K.; Punishing and abstaining for public goods, *Proceedings of the National Academy of Sciences of the United States of America* **103**, 495–497, 2006.

Brede, M.; Short versus long term benefits and the evolution of cooperation in the prisoner's dilemma game, *PLOS One*, 0056016, 2013.

Celiker, H., Gore, J.; Clustering in community structure across replicate ecosystems following a long-term bacterial evolution experiment, *Nature Communications* **5**, 4643, 2014.

Chalub, F. A. C. C., Santos, F. C., Pacheco, J. M.; The evolution of norms, *Journal of Theoretical Biology* **241**, 233–240, 2006.

Chen, X.-J., Wang, L.; Effects of cost threshold and noise in spatial snowdrift games with fixed multi-person interactions, *EPL* **90 (3)**, 38003, 2010.

Cui, P.-B., Wu, Z.-X.; Selfish punishment with avoiding mechanism can alleviate both first-order and second-order social dilemma, *Journal of Theoretical Biology* **361**, 111–123, 2014.

Dercole, F., De Carli, M., Della Rossa, F., Papadopoulos, A.V.; Overpunishing is not necessary to fix cooperation in voluntary public goods games, *Journal of Theoretical Biology* **326**, 70–81, 2013.

Doebeli, M., Hauert, C.; Models of cooperation based on the Prisoner's Dilemma and the Snowdrift game, *Ecology Letters* **8**, 748–766, 2005.

Dreber, A., Rand, D.G., Fudenberg, D., Nowak, M.A.; Winners don't punish, *Nature* **452**, 348–351, 2008.

Driscoll, W. W., Pepper, J. W.; Theory for the evolution of diffusible external goods, *Evolution* **64**, 2682–2687, 2010.

Du, F., Fu, F.; Quantifying the impact of noise on macroscopic organization of cooperation in spatial games, *Chaos, Soliton & Fractals* **56,** 35–44, 2013.

Ebel, H. and Bornholdt, S.; Coevolutionary games on networks, *Physical Review E.* 66, 056118, 2002.

EL-Seidy, E.; The effect of noise average relatedness between players in iterated games, *Applied Mathematics and Computation* **269**, 343–350, 2015.

Ermentrout, G. B., Edelstein, K. L.; Cellular Automata Approaches to Biological Modeling, *Journal of Theoretical Biology* **160**, 97–133, 1993.

Fehl, K., van der Post, D. J., Semmann, D.; Co-evolution of behavior and social network structure promotes human cooperation, *Ecology Letters* **14**, 546–551, 2011.

Fehr, E., Fischbacher, U.; The nature of human altruism, *Nature* **425**, 785–791, 2003.

Fehr, E., Gächter, S.; Altruistic punishment in humans, *Nature* **415**, 137–140, 2001.

Fehr, E., Gächter, S.; Altruistic punishment in humans, *Nature* **415**, 137–140, 2002.

Fowler, J.H.; Altruistic punishment and the origin of cooperation, *Proceedings of the National Academy of Sciences of the United States of America* **102**, 7047–7049, 2005.

Fukuda, E., Tanimoto, J.; Effects of stubborn decision-makers on vaccination and disease propagation in social networks, *International Journal of Automation and Logistics* **2**, 78–92, 2016.

Gächter, S., Renner, E., Sefton, M.; The Long-run benefits of punishment, *Science* **322**, 1510, 2008.

Gao, S., Wu, T., Nie, S., Wang, L.; Promote or hinder? The role of punishment in the emergence of cooperation, *Journal of Theoretical Biology* **386**, 69–77, 2015.

Garcia, J., Traulsen, A.; Leaving the loners alone Evolution of cooperation in the presence of antisocial punishment、 *Journal of Theoretical Biology* **307**, 168–173, 2012.

Garcia, J., van Veelen, M., Traulsen, A.; Evil green beards: Tag recognition can also be used to withhold cooperation in structured populations, *Journal of Theoretical Biology* **360**, 181–186, 2014.

Gardner, A., West, S.A.; Cooperation and punishment, especially in humans, *The American Naturalist* **164**, 753–764, 2004.

Gore, J., Youk, H., Van Oudenaarden, A.; Snowdrift game dynamics and facultative cheating in yeast, *Nature* **459**, 253–256, 2009.

Greig, D., Travisano, M.; The Prisoner's Dilemma and polymorphism in yeast SUC genes, *Proceedings of the Royal Society of London Series B: Biological Sciences* **271**, 25–26, 2004.

Gurerk, O., Irlenbusch, B., Rockenbach, B.; The competitive advantage of sanctioning institutions, *Science* **312**, 108–111, 2006.

Hamilton, W.D.; The evolution of altruistic behavior, *American Naturalist* **97**, 354–356, 1963.

Hamilton, W. D.; The genetical evolution of social behavior, *Journal of Theoretical Biology.* 7, 1–16, 1964.

Hao, D., Rong, Z., Zhou, T.: Extortion under uncertainty: Zero-dominant strategies in noisy games, *Physical Review E* **91**, 052803, 2015.

Hardin, G.; The Tragedy of the Commons, *Science* **162**, 1243–1248, 1968.

Hassell, M. P., Comins, H. N., and May, R. M.; Species coexistence and self-organizing spatial dynamics, *Nature* 313, 10–11, 1994.

Hauang, K., Wang, T., Cheng, Y., Zheng, X.; Effect of heterogeneous investments on the evolution of cooperation in spatial public goods game, *PLOS One*, 0120317, 2015.

Hauert, C.; Effects of space in 2×2 games, *International Journal of Bifurcation and Chaos* **12**, 1531–1548, 2002.

Hauert, C., De Monte, S., Hofbauer, J., Sigmund, K.; Volunteering as Red Queen mechanism for cooperation in public goods games, *Science* **296 (5570)**, 1129–1132, 2002.

Helbing, D., Yu, W.; The outbreak of cooperation among success-driven individuals under noisy conditions, *Proceedings of the National Academy of Sciences of the United States of America* **106 (10)**, 3680–3685, 2009.

Helbing, D., Szolnoki, A., Perc, M., Szabó, G.; Evolutionary establishment of moral and double moral standards through spatial interactions, *PLoS Computational Biology* **6**, e1000758, 2010.

Henrich, J., Boyd, R., Bowles, S., Camerer, C., Fehr, E., Gintis, H., McElreath, R.; In search of homo economicus: behavioral experiments in 15 small-scale societies, *The American Economic Review* **91**, 73–78, 2001.

Jeong, H.-C., Oh, S.-Y., Allen, B., Nowak, M. A.; Optional games on cycles and complete graphs, *Journal of Theoretical Biology* **356**, 98–112, 2014.

Julou, T., Mora, T., Guillon, L., Croquette, V., Schalk, I. J., Bensimon, D., Desprat, N.; Cell–cell contacts confine public goods diffusion inside Pseudomonas aeruginosa clonal microcolonies, *Proceedings of the National Academy of Science of the United States of America* **110**, 12577–12582, 2013.

Kokubo, S., Wang, Z., Tanimoto, J.; Spatial reciprocity for discrete, continuous and mixed strategy setups, *Applied Mathematics and Computation* **259**, 562–568, 2015.

Koschwanez, J. H., Foster, K. R., Murray, A. W.; Improved use of a public good selects for the evolution of undifferentiated multicellularity, *eLIFE* **2**, e00367, 2013.

Kraines, D., Kraines, V.; Pavlov and the Prisoner's Dilemma, *Theory Decision* **26**, 47–79, 1988.

Leimar, O., Hammerstein, P; Evolution of cooperation through indirect reciprocity, P*roceedings of the Royal Society B* **268**, 745–753, 2001.

Maire, T., Youk, H.; Molecular-Level Tuning of Cellular Autonomy Controls the Collective Behaviors of Cell Populations, *Cell Systems* **1**, 349–360, 2015.

Masuda, N.; Evolution of cooperation driven by zealots, *Scientific Reports* **2**, 646, (2012).

Matsuzawa, R., Tanimoto, J., Fukuda, E.; Spatial prisoner's dilemma games with zealous cooperators, *Physical Review E* **94**, 022114, 2016.

Maynard Smith, J.; Group selection, *Quarterly Review of Biology*. 51, 277–283, 1976.

Michor, F., Nowak, M. A.; The good, the bad and the lonely, *Nature* **419**, 677–678, 2002.

Milinski, M., Rockenbach, B.; Human behaviour: Punisher pays, *Nature* **452**, 297–298, 2008.

Miyaji, K., Wang, Z., Tanimoto, J., Hagishima, A., Ikegaya, N.; Direct reciprocity in spatial populations enhances R-reciprocity as well as ST-Reciprocity, *PLOS One* **8 (8)**, e71961, 2013.

Momeni, B., Waite, A. J., Shou, W.; Spatial self-organization favors heterotypic cooperation over cheating, *eLife* **2**, e00960, 2013.

Nakamaru, M., Iwasa, Y.; The evolution of altruism by costly punishment in lattice-structured populations: score-dependent viability versus score-dependent fertility, *Evolutionary Ecology Research* **7**, 853–870, 2005.

Nowak, M.A.; Five Rules for the evolution of cooperation, *Science* **314**, 1560–1563, 2006.

Nowak, M. A. and May, R. M.; Evolutionary games and spatial chaos, *Nature* 359, 826–829, 1992.

Nowak, M. A. and Sigmund, K.; Evolution of indirect reciprocity by image scoring, *Nature* 393, 573–577, 1998

Nowak, M. A., Sigmund, K.; Evolution of indirect reciprocity, *Nature* **337**, 1291–1298, 2005

Ohtsuki, H., Iwasa, Y.; How should we define goodness? – reputation dynamics in indirect reciprocity, *Journal of Theoretical Biology* **231**, 107–120, 2004.

Ohtsuki, H., Iwasa, Y.; The leading eight: social norms that can maintain cooperation by indirect reciprocity, *Journal of Theoretical Biology* **239**, 435–444, 2006.

Ohtsuki, H. and Nowak, M.A.; The replicator equation on graphs, *Journal of Theoretical Biology* 243, 86–97, 2006.

Ohtsuki, H., Hauert, C., Lieberman, E., and Nowak, M. A.; A simple rule for the evolution of cooperation on graphs and social networks, *Nature* 441, 502–505, 2006.

Olejarz, J. W., Nowak, M. A.; Evolution of staying together in the context of diffusible public goods, *Journal of Theoretical Biology* **360**, 1–12, 2014.

Pachanathan, K., Boyd, R.; A tale of two defectors: the importance of standing for evolution of indirect reciprocity, *Journal of Theoretical Biology* **224**, 115–126, 2003.

Pacheco, J. M., Santos, F. C., Chalub, F. A. C.; Stern-judging: A simple, successful norm which promotes cooperation under indirect reciprocity, *PLoS Computational Biology* **2**, e178, 2006.

Panchanathan, K., Boyd, R., Indirect reciprocity can stabilize cooperation without the second-order free rider problem, *Nature* **432**, 499–502, 2004.

Perc, M.; Double resonance in cooperation induced by noise and network variation for an evolutionary Prisoner's Dilemma, *New Journal of Physics* **8**, 183, 2006a.
Perc, M.; Chaos promotes cooperation in the spatial prisoner's dilemma game, *EPL* **75 (6)**, 841–846, 2006b
Perc, M.; Uncertainties facilitate aggressive behavior in a spatial Hawk-Dove Game, *International Journal of Bifurcation and Chaos* **17 (11)**, 4223–4227,2007.
Perc, M.; Coevolutionary games—A mini review, *Biosystems* **99** (2), 109–125, 2010.
Perc, M., Marhl, M., Evolution and dynamical coherence resonance in the pair approximated prisoner's dilemma game, *New Journal of Physics* **8**, 142, 2006.
Perc, M., Gomez-Gardenes, J., Szolnoki, A., Floria, L.M., Moreno, Y.; Evolutionary dynamics of group interactions on structured populations; a review, *Journal of the Royal Society Interface* **10**, 20120997, 2013.
Rand, D. G., Ohtsuki, H., Nowak, M. A.; Direct reciprocity with only costly punishment: Generous tit-for-tat prevails, *Journal of Theoretical Biology* **256**, 45–57, 2009.
Rand, D. G., Armao IV, J. J., Nakamaru, M., Ohtsuki, H.; Anti-social punishment can prevent the co-evolution of punishment and cooperation, *Journal of Theoretical Biology* **265**, 624–632, 2010.
Riolo, R., Cohen, M. D., Axelod, R.; Evolution of cooperation without reciprocity, *Nature* **414**, 441–443, 2001.
Roberts, G., Sherratt, T. N.; Does similarity breed cooperation?, *Nature* **418**, 499–500, 2002.
Rockenbach, B., Milinski, M.; Game theory: How to treat those of ill repute, *Nature* **457**, 39–40, 2009.
Santos, F. C. and Pacheco, J. M.; Scale-free networks provide a unifying framework for the emergence of cooperation, *Physical Review Letter*. 95, 098104, 2005.
Santos, F. C., Pacheco, J. M., and Lenaerts, T.; Cooperation prevails when individuals adjust their social ties, *PLOS Computational Biology* 2, 1284–1291, 2006.
Sasaki, T., Brännström, A., Dieckmann, U., Sigmund, K.; The take-it-or-leave-it option allows small penalties to overcome social dilemmas, *Proceedings of the National Academy of Sciences of the United States of America* **109**, 1165–1169, 2012.
Schoenmakers, S., Hilbe, C., Blasius, S. Traulsen, A.; Sanctions at honest signals The evolution of pool punishment by public sanctioning institutions, *Journal of Theoretical Biology* **356**, 36–46, 2014.
Schuessler, R.; Exit Threats and Cooperation under Anonymity, *Journal of Conflict Resolution* **33**, 728–749, 1989.
Sekiguchi, T., Nakamaru, M.; Effect of the presence of empty sites on the evolution of cooperation by costly punishment in spatial games, *Journal of Theoretical Biology* **256**, 297–304, 2009.
Shen, C., Lu, J., Shi, L.; Does coevolution setup promote cooperation in spatial prisoner's dilemma game?, *Applied Mathematics and Computation* **290**, 201–207, 2016.
Shinao, H., Nakamaru, M.; Strict or graduated punishment? Effect of punishment strictness on the evolution of cooperation in continuous Public Goods Game, *PLOS One* **8 (3)**, e59894, 2013.
Sigmund, K., Hauert, C., Nowak, M.A.; Reward and punishment, *Proceedings of the National Academy of Sciences of the United States of America* **98**, 10757–10762 , 2001.
Sigmund, K., Silva, H.D., Traulsen, A., Hauert, C.; Social learning promotes institutions for governing the commons, *Nature* **466**, 861–863, 2010.
Slatkin, M. and Wade, M. J.; Group selection on a quantitative character, *Proceedings of the National Academy of Science of the United States of America* 75, 3531–3534, 1978.
Sugden, R., Economics of right, Cooperation and welfare, Basil Blackwell, 1986.
Szabo, G., Fath, G.; Evolutionary games on graphs, *Physics Reports* **446**, 97–216, 2007.
Szolnoki, A., Perc, M., Szabo, G.; Topology independent impact of noise on cooperation in spatial public goods games, *Physical Review E* **80**, 056109, 2009a.
Szolnoki, A., Vukov, I., Szabó, G.; Selection of noise level in strategy adoption for spatial social dilemmas, *Physical Review E* **80**, 056112, 2009b.

Szolnoki, A., Szabó, G., Czakó, L.; Competition of individual and institutional punishments in spatial public goods games, *Physical Review E* **84**, 046106, 2011a.

Szolnoki, A., Szabo, G., Perc, M.; Phase diagram for the spatial public goods game with pool punishment, *Physical Review E* **83**, 036101, 2011b.

Szolnoki, A., Mobilia, M., Jiang, L.-L., Szczensny, B., Rucklidge, A. M., Perc, M.; Cyclic dominance in evolutionary games: a review, *Journal of the Royal Society Interface* **11**, 20140735, 2014.

Tanimoto, J; Promotion of cooperation by payoff noise in a 2 × 2 game, *Physical Review E* **76**, 041130, 2007.

Tanimoto, J.; A simple scaling of the effectiveness of supporting mutual cooperation in donor-recipient games by various reciprocity mechanisms, *BioSystems* **96**, 29–34, 2009.

Tanimoto, J.; Mathematical Analysis of Environmental System, Springer (Tokyo), 2014.

Tanimoto, J.; Fundamentals of Evolutionary Game Theory and its Applications, Springer (Tokyo), 2015.

Tanimoto, J., Ogasawara, T.; Dynamic noise from action errors enhances network reciprocity in the prisoner's dilemma game, *Journal of Statistical Mechanics: Theory and Experiment*, P1033, 2015.

Tanimoto, J., Sagara, H.: Relationship between dilemma occurrence and the existence of a weakly dominant strategy in a two-player symmetric game, *BioSystems* **90 (1)**, 105–114, 2007.

Tanimoto, J., Sagara, H.; How the indirect reciprocity with co-evolving norm and strategy for 2 × 2 prisoner's dilemma game works for emerging cooperation, *Physica A* **438**, 595–602, 2015.

Taylor, M. and Nowak, M. A.; Transforming the dilemma, *Evolution* **61 (10)**, 2281–2292, 2007.

Traulsen, A., Claussen, C.; Similarity- based cooperation and spatial segregation, *Physical Review E* **70**, 046128, 2004.

Traulsen, A. and Nowak, M. A.; Evolution of cooperation by multilevel selection, *Proceedings of the National Academy of Science of the United States of Americ* 103, 10952–10955, 2006.

Trivers, R.; The evolution of reciprocal altruism, *Quarterly Review of Biology* **46**, 35–37, 1971.

Trivers, R.; Social Evolution, Benjamin/Cummings, Menlo Park. CA. 1985.

Velicer, G. J.; Social strife in the microbial world, *Trends in Microbiology* **11**, 330–337, 2003.

Wang, Z., Kokubo, S., Tanimoto, J, Fukuda, E., Shigaki, K.; Insight on the so-called spatial reciprocity, *Physical Review E* **88**, 042145, 2013.

Wang, Z., Kokubo, S., Jusup, M, Tanimoto, J.; Universal scaling for the dilemma strength in evolutionary games, *Physics of Life Reviews* **14**, 1–30, 2015a.

Wang, Z., Wang, L., Szolnoki, A., Perc, M.; Evolutionary games on multilayer networks: a colloquium, *European Physical Journal B* **88 (124)**, 60270–7, 2015b.

Wang, Q., He, N., Chen, X.; Replicator dynamics for public goods game with resource allocation on large population, *Applied Mathematics and Computation* **328**, 162–170, 2018.

Wedeking, C., Milinski, M.; Cooperation through image scoring in humans, *Science* **288**, 850–852, 2000.

West, S. A., Diggle, S. P., Buckling, A., Gandner, A., Griffin, A. S.; The social lives of microbes, *Annual Review of Ecology, Evolution, and Systematics* **38**, 53–77, 2007.

Williams, G. C.; Adaption and natural selection: A critique of some current evolutionary thought, Princeton Univ. Press, Princeton. NJ., 1996.

Wilson, D. S.; A theory of group selection, *Proceedings of the National Academy of Science of the United States of America* 72, 143–146, 1975.

Wynne-Edwards, V. C.; Animal dispersion in relation to social behavior, Oliver and Boyd, Edinburg, 1962.

Yamauchi, A., Tanimoto, J., and Hagishima, A.; What controls network reciprocity in the prisoner's dilemma game? *BioSystems* 102(2–3), pp.82–87, 2010.

Yao, Y., Chen, S.-S.; Multiplicative noise enhances spatial reciprocity, *Physica A* **413**, 432–437, 2014.

Yokoi, H., Uehara, T., Sakata, T., Naito, H., Morita, S., Tainaka, K.; Evolution of altruism in spatial prisoner's dilemma: Intra- and inter-cellular interactions, *Physica* A **416**, 361–370, 2014.

Zhang, G.-Q., Sun, Q.-B., Wang, L.; Noise-induced enhancement of network reciprocity in social dilemmas, *Chaos, Solitons & Fractals* **51**, 31–35, 2013.

Zhang, G.-Q., Hu, T.-P., Yu, Z.; An improved fitness evaluation mechanism with noise in prisoner's dilemma game, *Applied Mathematics and Computation* **276**, 31–36, 2016.

Zhong, W., Kokubo, S., Tanimoto, J.; How is the equilibrium of continuous strategy game different from that of discrete strategy game?, *BioSystems* **107 (2)**, 89–94, 2012.

Chapter 3
Social Dilemma Analysis for Modeling Traffic Flow

In this chapter, we concern ourselves with traffic flow as an example of an intrinsically dynamical system to which evolutionary game theory can be applied. The study of traffic flow was originally thought to be best explained using fluid dynamics in the mid-twentieth century, mainly in the field of civil engineering, owing to the recognition of traffic jams as a pressing urban problem. In the 1990s, complexity science, strongly supported by the rapid growth of computational resources, re-highlighted traffic-flow analysis with the concept of multi-agent simulation (MAS), allowing researchers to incorporate microscopic events, such as merging and branching lanes, lane-changing, and drivers' decision-making characteristics, into simulations to better reproduce real traffic flows.

The chapter begins by discussing observation of traffic flow from two points of view: macroscopic and microscopic concepts. As mentioned above, the former comes from macroscopic fluid dynamics, while the latter results from the microscopic idea of how each vehicle moves. Amazingly, or as might be expected, it has been proved that those specific two ideas are mathematically equivalent. Meanwhile, for practical applications, cellular-automaton (CA) simulation—a specific form of MAS containing both an agent-action rule and time-evolutionary rule—has been most heavily used for modeling because of its flexibility and the transparency of its physics. After discussing one upcoming MAS model, we see the practical two examples in which traffic-flow analysis intriguingly dovetails with evolutionary game theory: the rote-selection problem and the 2-body dynamics of each vehicle trying to change lanes while hampering others' lane-changes. These issues show us that social dilemmas are strongly behind the working of traffic flows. Inversely speaking, traffic-flow systems impose multiplayer-dilemma structures on all drivers. Consequently, traffic jams are inevitable when public social efficiency is reduced.

J. Tanimoto, *Evolutionary Games with Sociophysics*, Evolutionary Economics and Social Complexity Science 17, https://doi.org/10.1007/978-981-13-2769-8_3

3.1 Traffic-Flow Analysis; Microscopic and Macroscopic Approaches

It would be quite natural to think of traffic flow as being analogous to a pipe flow (Fig. 3.1). Abstracting all details, both seem to be one-dimensional systems relating to fluid dynamics. However, there is an important difference: the transferred body in an ordinary pipe flow is continuous—e.g., water, oil, or humid air—whereas, bodies in a traffic flow are discrete.[1] Depending on your preference for physics jargon, each vehicle should be called a particle or granular agent. In fluid dynamics, two opposing viewpoints can be adopted: Eulerian-scope and Lagrangain-scope. As a standard textbook explains, the Lagrangian-scope relies on a set of governing equations from the Newton's secondary law describing a mass-point dynamics, whereas the set of governing equations for Eulerian-scope describes a bulk balance of either momentum, energy, or scalars for an infinitely small volume, and does not discuss each particle's dynamics at all. The Eulerian and Lagrangian scopes respectively provide foundations for the macroscopic and microscopic features of the fluid.

Although not apparent from Fig. 3.1, there is another, arguably more important, difference between pipe flow and traffic flow: namely, whether the incompressible-fluid approximation is applicable. In the case of ordinal pipe flows, unless the flow

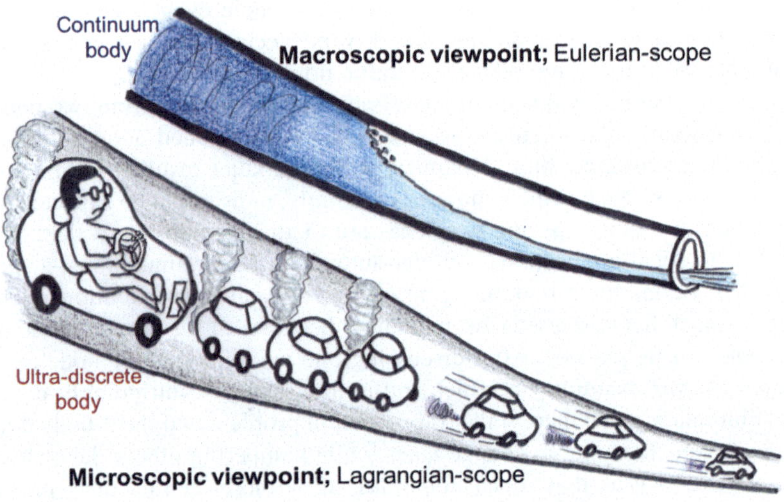

Fig. 3.1 Macroscopic versus microscopic concepts for observing flows

[1]Precisely speaking, it should be called "ultra-discrete", as shown in Fig. 3.1. According to the convention in the fields of mathematics and physics, the term "discrete" has been used for digitalization for either 'time' or 'space'. In the current discussion, the object to be digitalized is the body to be transferred.

velocity is extremely high, the flow field can be approximated as incompressible. On the contrary, traffic flow is compressible. In fact, its local density, as determined by both local flux and local velocity, is obviously changeable in time. Because of this feature, a gam takes place.

Returning to the first point, we should note that the difference between continuous and ultra-discrete is relative. Assuming an extremely fine resolution, water and air can be seen as molecular flows with an immensely large number of particles. This suggest that traffic flow is formulated by a set of equations for 1D (one-dimensional) fluid flow. It is the macroscopic concept for traffic-flow analysis.

3.1.1 Macroscopic Concept[2]

Imagine an ideal 1D traffic-flow system with a single lane, and let us define local traffic density ρ [vehicles/m], local velocity v [m/s], and local flux q [vehicles/s] at a certain infinitely small road section. Time and space are denoted by t and x, respectively. Note the following relation among these three variables:

$$q = \rho \cdot v. \tag{3.1}$$

This relation can be justified by dimensional considerations. In a sense, this gives the definition of flux, along with the given traffic density and velocity.

From the "conservation law" appearing in any standard textbook on fluid dynamics, we can draw the following non-linear first-order partial differential equation, which describes conservation of vehicles for an infinitely small road section:

$$\frac{\partial \rho}{\partial t} + \frac{\partial q}{\partial \rho} \cdot \frac{\partial \rho}{\partial x} = 0. \tag{3.2}$$

Equation (3.2) obviously presumes a compressive-fluid system.

The set of Eqs. (3.1) and (3.2) is not solvable because the number of unknown variables, 3, is larger than the number of equations, 2. Paraphrasing, we need another equation giving a non-trivial relation among the three variables to close the set of governing equations.

One likely scenario is that the relationship between equilibrium local density and velocity,

$$v = v(\rho), \tag{3.3}$$

[2]With respect to the macroscopic concept, readers can consult the following text book: Haberman (1977).

may be given. Noting that the inverse of local traffic density physically means the local average inter-vehicular distance, we may fix an exact form of Eq. (3.3) from field-measurement datasets. Or, field measurements may tell us

$$q = q(\rho). \tag{3.4}$$

Equation (3.4) is mathematically equivalent to Eq. (3.3). Taking the first derivative with respect to ρ for Eq. (3.4), we recover Eq. (3.3). Thus, giving the relation between local density and velocity is equivalent to giving the relation between local density and flux.

One crucial point of discussion is whether we can fix either Eq. (3.3) or (3.4), e.g., whether one single velocity (or flux) can be determined whenever a traffic density is given. In an ideal and imaginary traffic flow, this would be possible. In such a case, the set of governing equations gives us a solution, because the closure problem of the governing equations is fixed. However, in the case of realistic traffic flow, it is impossible. Figure 3.2 presents the $q - \rho$ relationship, called the "fundamental diagram", obtained for a certain Japanese highway.[3] The plots in the range of lower traffic density show in a linear clump, because a vehicle is able to move at a maximum velocity expressed as $\frac{\mathrm{d}q(\rho)}{\mathrm{d}\rho}$, with little mutual interference; this is called the free-flow phase. The plots in middle density, highlighted by the yellow triangle,[4] show extremely scattered. In particular, for $0.1 < \rho < 0.15$, although some data show quite high traffic flux, fitting to the gray line to sum up the free-flow phase, which is called meta-stable phase, there are some plots indicating small flux down to around $q = 0.2$. This results from the fact that, in the meta-stable phase, a series of vehicles moves over small inter-vehicular distances, known as "platoon driving". Thus, given a small

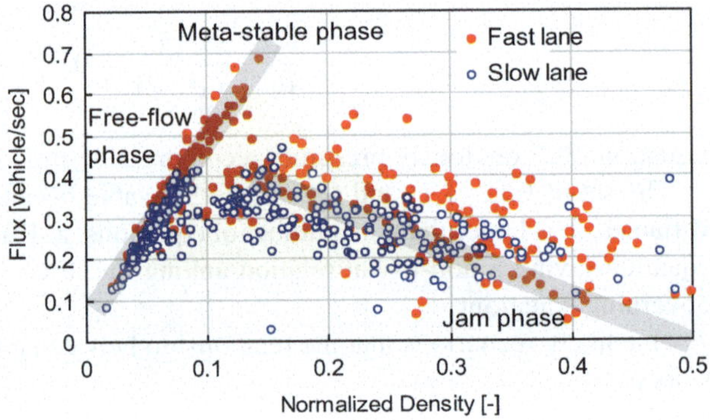

Fig. 3.2 Fundamental diagram obtained by a field-measurement campaign on Japanese highways (see Tanimoto et al.[3]). Blue and red plots respectively show data (for overtaking) in the slow and fast lanes in the two-lane system

[3]Tanimoto et al. (2015).

[4]Kerner called this plot the chattered area, with the region sandwiched between the free-flow phase and jam phase called the 'synchronized phase' He insisted that a real traffic-flow field must contain those three phases, and a phase transition from free to jam takes place via the synchronized phase (not F- > J directly but F- > S- > J). This is called Kerner's three-phase theory; for details, readers can consult. Kerner (2009).

perturbation resulting from one of the vehicles unintentionally braking, for example, external noise, the platoon of vehicles would collapse into a situation where certain vehicles move with a lower velocity. They never return to the original platoon state spontaneously. Because of this irreversible process, statistical physicists terms this phenomenon a 'meta-stable' phase. Considering this, either Eq. (3.3) or Eq. (3.4) is underspecified. If this is the case, the set of governing equations is not closed.

One idea for closing the set of equations is to describe how each vehicle accelerates or decelerates. But this idea cannot be microscopic, even though it tries to evaluate how each vehicle moves. This idea comes from the optimal-velocity (OV) model, a major microscopic model; it uses another macro equation to describes how a vehicle adjusts his own velocity (by acceleration or deceleration) to the optimal velocity defined by the local density (or local average inter-vehicular distance), $V_e(\rho)$, in an average relaxation time τ:

$$\frac{\partial v}{\partial t} + v\frac{\partial v}{\partial x} = \frac{V_e(\rho) - v}{\tau} + (\text{additional terms}). \tag{3.5}$$

Qualitatively, $V_e(\rho)$ approaches a maximum velocity at the lower limit of density, but goes to zero with $\rho \to 1$. The OV function, $V_e(\rho)$, can be determined by field-measurement-data sets. The additional terms in Eq. (3.5) indicate other contributions to acceleration or deceleration. There have been various models related to this. For example, Payne proposed[5]

$$\frac{\partial v}{\partial t} + v\frac{\partial v}{\partial x} = \frac{V_e(\rho) - v}{\tau} + \frac{1}{\rho\tau} \cdot \frac{V_e'(\rho)}{2} \cdot \frac{\partial \rho}{\partial x}. \tag{3.6}$$

Jiang, considering traffic sound speed, C, gave[6]

$$\frac{\partial v}{\partial t} + v\frac{\partial v}{\partial x} = \frac{V_e(\rho) - v}{\tau} + C \cdot \frac{\partial v}{\partial x}. \tag{3.7}$$

Equations (3.5), (3.6), and (3.7) share same mathematical form, which is analogous to the 1D momentum-transport equation for conventional fluid dynamics. Applying one of those here closes the set of governing equations.

3.1.2 Microscopic Concept

Unlike the continuum model backed by the macroscopic concept, the microscopic concept considers each vehicle; its formulation therefore comes from Newton's

[5] Payne (1971).

[6] Jiang et al. (2002).

second law. A macroscopic-continuum model utilizes an analytic approach to obtain a solution, although a numerical procedure may be applied in the final stage. In the class of the microscopic concept, two approaches—analysis and simulation—are possible depending on what type of model is presumed. For instance, as we will discuss later, a model that uses cellular automata (CA) to consider either the detailed actions of drivers or the large range of driver diversity in a real traffic flow can be used as a simulation approach.

Below, we discuss some representative models based on the microscopic concept

3.1.2.1 Optimal-Velocity Model

Assuming a Lagrangian scope, Bando et al. posed the following optimal-velocity (OV) model,[7] which, rather than Eq. (3.5), is formulated as

$$\frac{dv_i(t)}{dt} = a[V_e(\Delta x_i(t)) - v_i(t)], \tag{3.8}$$

where $x_i(t)$ is the position of vehicle i at time t, $\Delta x_i(t) = x_{i+1}(t) - x_i(t)$ is the headway distance, a is the sensitivity of a driver that can be rewritten as $1/\tau$, appearing in Eq. (3.5), and V_e is the OV function adopted by Bando et al.:

$$V_e(\Delta x) = \frac{V_{\max}}{2}[\tanh(\Delta x - x_c) + \tanh(x_c)]. \tag{3.9}$$

Here, x_c is the infection point of the OV function giving $V_e = V_{\max}/2$. Figure 3.3 provides a sketch of the graph. The OV model presumes that headway distance primarily determines acceleration. Thus, it is called a "car-following model". Following this pioneering work by Bando et al., a hundred and one variant OV models have been proposed, each adding some new term(s) to modify the acceleration represented by the OV function. One representative model was proposed by Xue et al.[8]:

$$\frac{dv_i(t)}{dt} = a[V_e(\Delta x_n(t)) - v_i(t)] + \lambda \cdot \Delta v_i(t), \tag{3.10}$$

where λ is a different sensitivity coefficient than a and $\Delta v_i(t)$ is the velocity difference between the preceding vehicle $i + 1$ and the following (focal) one i at time t. This model insists that the acceleration/deceleration that a driver introduces is not attributable only to his attempts to reach the optimal velocity, but also to the velocity difference with the preceding vehicle.

---

[7] Bando et al. (1995).

[8] Xue (2002).

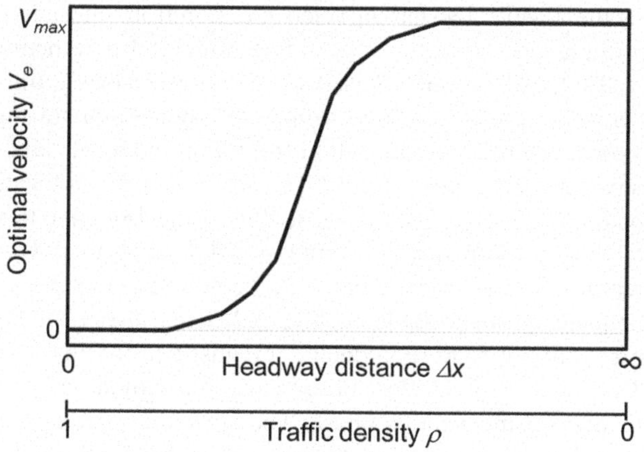

Fig. 3.3 Schematic example of an optimal-velocity (OV) function

OV and its variants presume that only one kind of equation is needed to describe each vehicle's motion. It does not require a conservation law like Eq. (3.2), because the microscopic view, which observes each of all vehicles in a system, basically assumes that no vehicles appear or disappear.

Like a continuum model, an analytic approach can be applied to an OV model to obtain a solution. In practice, the governing equation can be transformed into the modified KdV (m-KdV) equation, which brings an exact solution relating to the kink–antikink soliton.[9]

3.1.2.2 Cellular Automaton (CA) Model

Cellular-automaton (CA) models are discrete dynamical systems composed of calls presuming a set of simple local rules and a spatial structure such as a 2D lattice or a 1D ring.[10] The original concept was established in 1940' by Stanisław Marcin Ulam and John von Neumann. In 1980', Stephen Wolfram profoundly studied 1D CA (known as the elementary cellular automaton) and suggested that some CA classes are Turing-complete.

Let us consider the 1D CA Wolfram studied. Let us presume a simplest model, wherein the state of each cell is either 0 (vacant) or 1 (occupied) and this state is determined by the state of that particular cell and its neighbors (namely, three successive cells) in the previous time step. There are 256 rules in the set for the following reason: the number of states of the three cells is $2^3 = 8$, whereas the possible state of the focal cell in the next time step is binary, 0 or 1. Hence, the number of rules must be $2^8 = 256$. Wolfram explicitly described Rules-0 to -255.

[9]There have been many works on this subject; a representative one is: Peng et al. (2011).
[10]Wolfram (1983).

Of these, Rule-184 has attracted attention from physicists, because it seems to model a single lane of traffic well. In fact, Rule-184 is sometimes called the "traffic rule".

Table 3.1 presents Rule-184. It can reproduce a traffic flow because a vehicle, depicted by state '1', flows from left to right. Paraphrasing the rule, it demands that if a focal cell has a vacant cell in front of it, the focal cell can move 1 cell ahead in the next time step; otherwise it must stay at its present location. This seems simple and plausible for an idealized traffic flow. Rule-184 analytically presents the fundamental diagram, which is shown in Fig. 3.4. Comparing it with fundamental diagrams for real traffic, for instance Fig. 3.2, we note that Fig. 3.4 does not have a branch part, denoted as meta-stable in Fig. 3.2. This is one of the drawbacks of Rule-182, preventing it from reproducing a realistic traffic flow. To compensate, Takayasu & Takayasu[11] found that introducing **slow-to-start** (S2S) effect to a CA model recovers a meta-stable phase. The slow-to-start effect implies that a vehicle, once stopping, takes one more time-step to restart after a preceding cell becomes vacant. Another modification to the primitive Rule-184 CA model that should be mentioned here is the **quick-start** (QS) effect.[12] Quick-start considers that a vehicle can observe

Table 3.1 Wolfram's CA rule-184

Current state of successive 3 cells	State of the central cell in the next time step
111	1
110	0
101	1
100	1
011	1
010	0
001	0
000	0

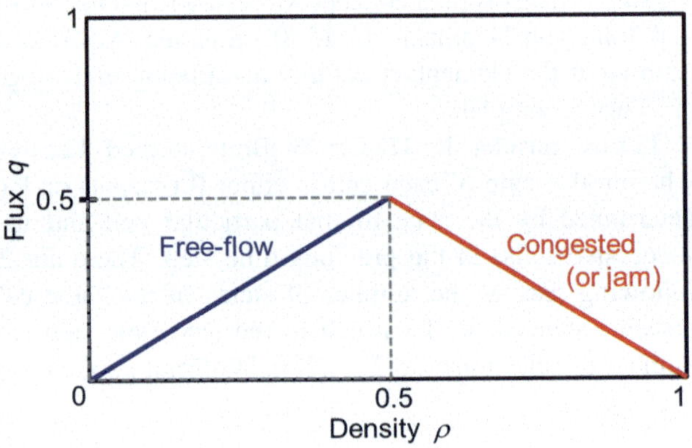

Fig. 3.4 Fundamental diagram ($\rho - q$ relation) by Wolfram's CA rule-184

[11]Takayasu and Takayasu (1993).

[12]Nishinari and Takahashim (2000).

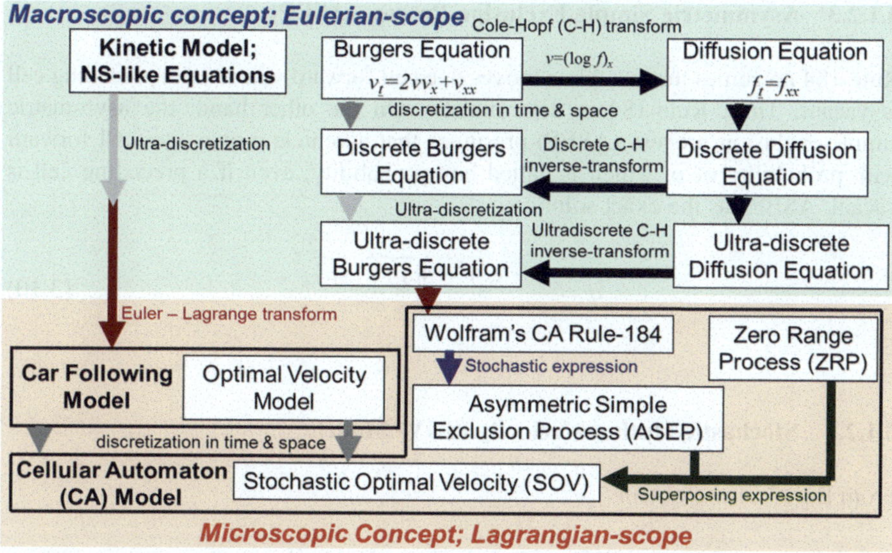

Fig. 3.5 Overview of traffic models: the macroscopic and microscopic approaches

further than the car immediately in front of it when deciding to accelerate or decelerate, which might be conceivable for observation of how drivers really adjust to the surrounding traffic-flow field. It is well-known that the introduction of quick-start to a simple CA model, for instance Rule-184, causes the point of peak flux (q (0.5) in Fig. 3.4) to shift to the right (discussed later), does not influence on emerging the branch of meta-stable at all. Sub-models for compensating the primitive CA models will be discussed later on (Fig. 3.5).

Perhaps one of the most surprising and important findings[13] relating to the CA model is that Rule-184 has been mathematically proven to be equivalent to one of the simplest continuum models, the Burgers equation, which means that a microscopic model is fully equivalent to a macroscopic one. This implies that microscopic and macroscopic concepts are mutually transformable. Burger's equation, one variant obeying the continuum model, can be expressed by $\frac{\partial v}{\partial t} + 2v\frac{\partial v}{\partial x} = \frac{\partial^2 v}{\partial x^2}$, which obviously belongs to the non-linear class. But by applying a Cole –Hoph (C – H) transform, it becomes a diffusion equation, one of the most well-known linear equations. It can be solved analytically. Moreover, by discretizing both time and space and applying an inverse-C – H transform, a discrete Burgers equation can be drawn. In addition, by applying **ultra-discretization**, the continuum velocity v can be transformed into a discrete variable indicating each vehicle. Hence, it comes to Rule-184.

[13]We suggest the following references: Nishinari and Takahashim (1998), Nishinari (2001), Tokihiro et al. (1996).

3.1.2.3 Asymmetric Simple Exclusion Process (ASEP)

Rule-184 presumes that a vehicle moves one cell forward whenever a preceding cell is vacant. Thus, Rule-184 is deterministic. On the other hand, the asymmetric simple-exclusion process (ASEP) presumes that a vehicle moves one cell forward with probability of p, which is called hop probability, even if a preceding cell is vacant. ASEP has the exact solution

$$q = \frac{1 - \sqrt{1 - 4p\rho(1-\rho)}}{2}. \tag{3.10}$$

3.1.2.4 Stochastic-Optimal-Velocity (SOV) Model

From Eq. (3.8), we obtain

$$v_i^{t+1} = (1-a)v_i^t + a \cdot V_e\left(\Delta x_i^t\right). \tag{3.11a}$$

We need the following rule to close this CA, namely the stochastic-optical-velocity (SOV) model:

$$x_i^{t+1} = x_i^t + v_i^{t+1}. \tag{3.11b}$$

One important aspect of this model is that v and V_e are normalized by V_{max} if $V_{max} > 1$. Hence, note that $0 \le v \le 1$ and $0 \le V_e(h) \le 1$ are valid, and that $0 \le a \le 1$ is needed. The normalized velocity, v, *is the* hop velocity.

Interestingly, SOV contains ASEP and a zero-range process (ZRP), where hop probability is given by a headway distance. In fact, presuming $a = 0$, ASEP ($v_i^{t+1} = v_i^t$) is recovered; and presuming $a = 1$, ZRP[14] ($v_i^{t+1} = V_e\left(\Delta x_i^t\right)$) is given. ASEP and ZRP are quite important CA models, because they have exact solutions.

3.1.2.5 Fukui – Ishibashi Model and Quick-Start Model

In this subsection, we prove the respective exact solutions of the Fukui – Ishibashi model[15] (hereafter, FI model) and the quick-start model[12] (hereafter QS model, briefly mentioned on the previous page).

The FI model, a CA model that can be solved exactly, presumes $V_{max} > 1$, e.g.,

---

[14] A reader can consult the following literature concerning ZRP and its exact solution. O'Loan et al. (1998).

[15] Fukui and Ishibashi (1996).

$$\text{if } \left(\Delta x_i^t - 1 \geq V_{\max}\right) \text{ then } \left(v_i^{t+1} = V_{\max}\right),$$
$$\text{if } \left(\Delta x_i^t - 1 < V_{\max}\right) \text{ then } \left(v_i^{t+1} = \Delta x_i^t - 1\right).$$

It is obvious from Fig. 3.6 that the critical density minimizing traffic flux is $\rho_{cri} = \frac{1}{V_{\max}+1}$. Thus, the maximum flux is drawn by $q_{\max} = \frac{V_{\max}}{V_{\max}+1}$. The exact solution and its fundamental diagram are shown below:

$$q = \frac{V_{\max}}{1+V_{\max}} - \left\{ \frac{1+V_{\max}}{2}\sqrt{\left(k - \frac{1}{1+V_{\max}}\right)^2} + \frac{1-V_{\max}}{2}\left(\rho - \frac{1}{1+V_{\max}}\right) \right\}. \tag{3.12}$$

The QS model is as follows:

Let S be the number of sites in front of a focal vehicle that are foreseen for quick-start; if they are empty, then the focal vehicle moves.

It is obvious from Fig. 3.7 that the critical density is $\rho_{cri} = \frac{S}{S+1}$. Thus, the maximum flux is drawn by $q_{\max} = \frac{S}{S+1}$, because $v = 1$. The exact solution and its fundamental diagram are shown below:

$$q = \frac{S}{S+1} - \left\{ \frac{S+1}{2}\sqrt{\left(\rho - \frac{S}{S+1}\right)^2} + \frac{S-1}{2}\left(\rho - \frac{S}{S+1}\right) \right\}. \tag{3.13}$$

[Proof 1]

As below, we prove that the fundamental diagrams of the FI (Fig. 3.8) and QS models (Fig. 3.9) can be described as an asymmetrical tent-type polygonal-line functions as below.

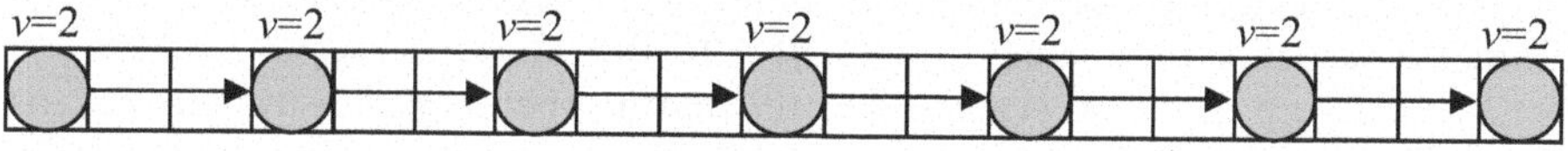

(a) $\rho = \rho_{cri}$ and V_{max}=2

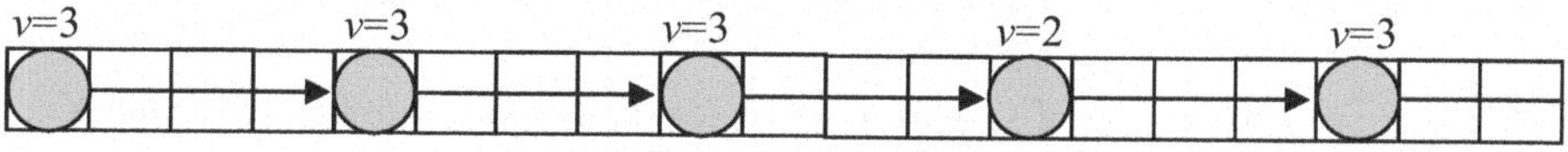

(b) $\rho = \rho_{cri}$ and V_{max}=3

Fig. 3.6 Flow states for the FI model

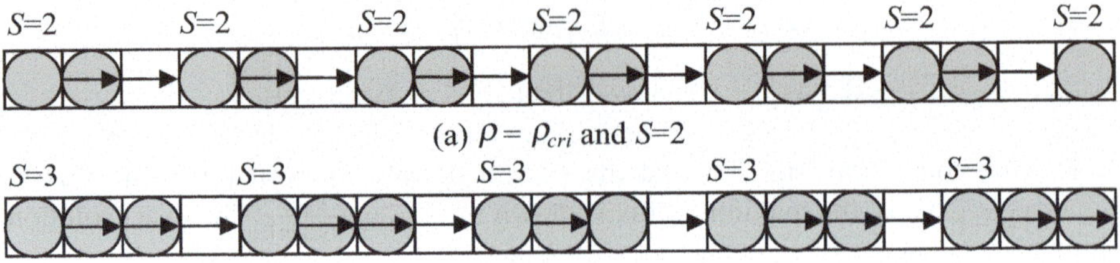

(b) $\rho = \rho_{cri}$ and S=3

Fig. 3.7 Flow states explaining for the QS model

Fig. 3.8 Fundamental diagram of the FI model obtained by Eq. (3.12)

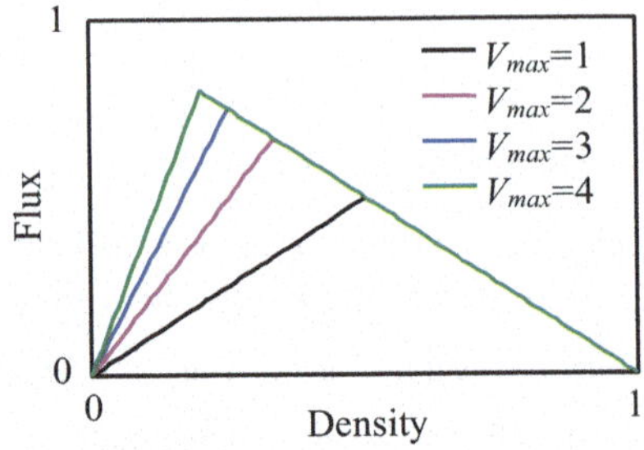

Fig. 3.9 Fundamental diagram of the QS model obtained by Eq. (3.13)

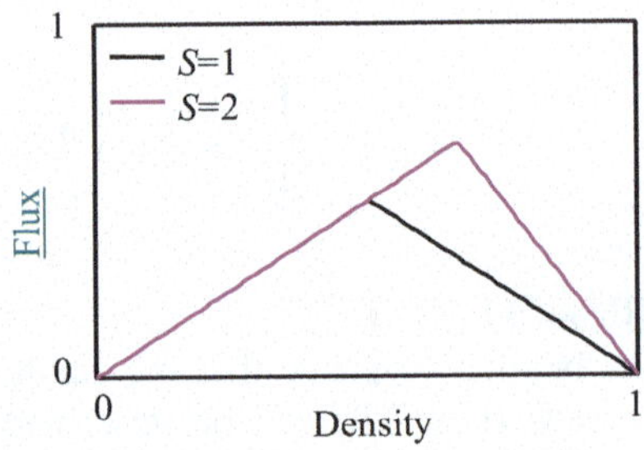

(i) FI model

When $\rho \leq \rho_{cri}$, it is trivial to show that the relation between density and flux must be a linear function originating from $(\rho, q) = (0, 0)$ with slope V_{max}.

When $\rho \geq \rho_{cri}$, let us evaluate the relationship between density and flux as follows. Figure 3.10 shows an example of the flow state when $\rho \geq \rho_{cri}$. From the schematic implication and the definition of average velocity, we know.

$$\text{(average velocity)} = \left(\text{total sum of all vehicles' velocity}\right) / \left(\text{number of vehicles}\right).$$

Trivially,

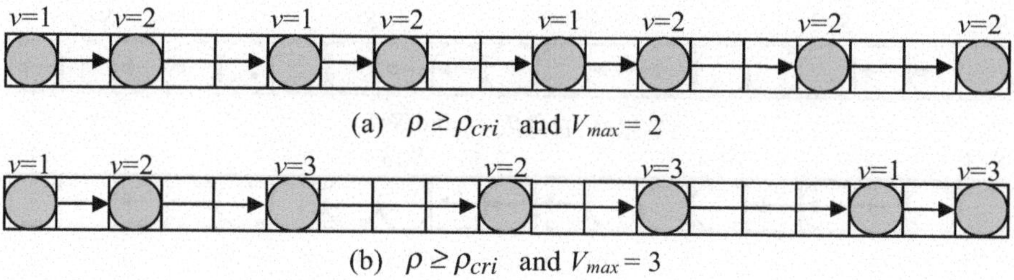

Fig. 3.10 Flow states for the FI model's proof

$$(\text{total sum of all vehicles' velocity}) = (1 - \rho)L,$$

where L is system length. Hence,

$$(\text{average velocity}) = L(1 - \rho)/(L\rho).$$

Therefore, flux $q(\rho)$ is.

$$q(\rho) = 1 - \rho,$$

which explicitly implies a linear function of ρ. This negative-slope line obviously crosses both the critical state and the complete-jam state, $\left(\frac{1}{V_{\max}+1}, \frac{V_{\max}}{V_{\max}+1}\right)$ and (1,0).

(QED)

(ii) QS model

When $\rho \leq \rho_{cri}$, it is trivial to show that the relation between density and flux must be a linear function originating from $(\rho, q) = (0, 0)$ with slope $V_{max.} = 1$.

When $\rho \geq \rho_{cri}$, let us evaluate the relationship between density and flux as follows. Figure 3.11 shows an example of a flow state when $\rho \geq \rho_{cri}$. From the schematic implication and definition of average velocity, we know that.

$$\begin{aligned}(\text{average velocity}) &= (\text{total sum of all vehicles' velocity})/(\text{number of vehicles}) \\ &= (\text{number of vehicles that can move})/(\text{number of vehicles}).\end{aligned}$$

Observing Fig. 3.11, we notice that vehicles staying in S sites behind each empty site can move. Thus.

$$(\text{number of vehicles that can move}) = L(1 - \rho)S,$$

and

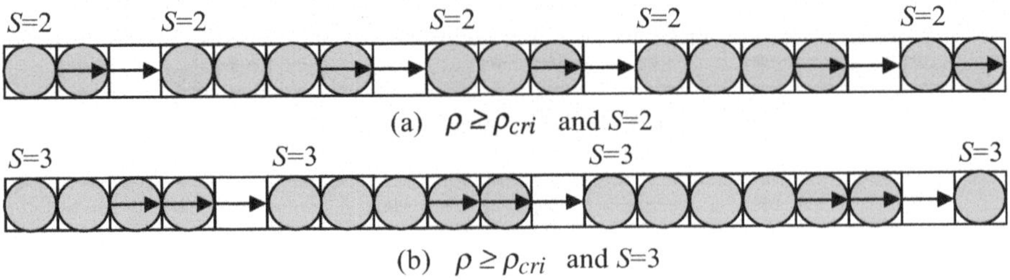

Fig. 3.11 Flow states for the QS model's proof

$$(\text{average velocity}) = L(1-\rho)S/(L\rho).$$

Therefore, flux $q(\rho)$ is.

$$q(\rho) = 1 - \rho,$$

which explicitly implies a linear function of ρ. This negative-slope line obviously crosses both the critical and complete-jam states, $\left(\frac{S}{S+1}, \frac{S}{S+1}\right)$ and (1,0).

(QED)

[Proof 2]

Below, we prove that Eqs. (3.12) & (3.13) can be drawn when we know only that it is an asymmetric-tent-type function crossing (0,0), (ρ_{cri}, q_{max}) and (1,0).

What we expect is that a function consisting of two lines can be expressed by a single expression: $f: Q = f(k)$(Fig. 3.12a). Shifting the function in the direction of k by $-p$, we obtain $f' : Q = f(k + p)$(Fig. 3.12b).

Meanwhile, two equations that touch at $(k, Q) = (0, q)$, f_1 and f_2 (as shown in Fig. 3.12b), can be expressed as below:

$$f_1 : Q = a\sqrt{k^2} + q, \tag{3.14}$$

$$f_2 : Q = bk + q. \tag{3.15}$$

The expression of f_1 is determined because it must be symmetric across the Q axis. By superposing those two linear functions, we obtain the asymmetric tent-type function

$$f' : Q = f(k + p) \equiv f_1 + f_2 - q.$$

By substituting Eqs. (3.14) & (3.15), we obtain:

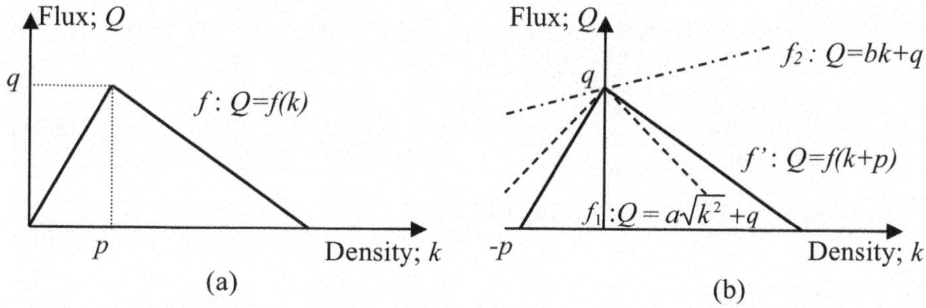

Fig. 3.12 Asymmetric tent-type function

$$f(k+p) = a\sqrt{k^2} + bk + q. \tag{3.16}$$

When the branches of the square root are considered, we have:

$$\text{if } \ k < 0 \text{ then}, f(k+p) = (-a+b)k + q \equiv D_1 k + q, \tag{3.17}$$

$$\text{if } \ k > 0 \text{ then}, f(k+p) = (a+b)k + q \equiv D_2 k + q. \tag{3.18}$$

(i) FI-model Case

We have $(p, q) = \left(\frac{1}{V_{\max}+1}, \frac{V_{\max}}{V_{\max}+1}\right)$ and $D_1 = V_{\max}, D_2 = -1$.
By substituting the above conditions into Eqs. (3.17) & (3.18), we obtain

$$a = -\frac{V_{\max}+1}{2}, b = \frac{V_{\max}-1}{2}. \tag{3.19}$$

By substituting these into Eq. (3.15), we find

$$f(k+p) = -\frac{V_{\max}+1}{2}\sqrt{k^2} + \frac{V_{\max}-1}{2}k + q. \tag{3.20}$$

As a consequence, *f*(*k*) is

$$f(k) = -\frac{V_{\max}+1}{2}\sqrt{(k-p)^2} + \frac{V_{\max}-1}{2}(k-p) + q.$$

Namely; we obtain;

$$\therefore Q = -\frac{V_{\max}+1}{2}\sqrt{\left(k-\frac{1}{V_{\max}+1}\right)^2}+\frac{V_{\max}-1}{2}\left(k-\frac{1}{V_{\max}+1}\right)$$
$$+\frac{V_{\max}}{V_{\max}+1}. \tag{3.21}$$

Evidently, Eq. (3.21) reads as Eq. (3.12).

(QED)

(ii) QS-model Case

We know that $(p,q)=\left(\frac{S}{S+1},\frac{S}{S+1}\right)$ and $D_1=1, D_2=-S$.
By substituting the above conditions into Eqs. (3.17) & (3.18), we obtain

$$a=-\frac{S+1}{2}, b=-\frac{S-1}{2}. \tag{3.22}$$

By substituting these into Eq. (3.15), we find

$$f(k+p)=-\frac{S+1}{2}\sqrt{k^2}-\frac{S-1}{2}k+q. \tag{3.23}$$

Therefore, *f*(*k*) is;

$$f(k)=-\frac{S+1}{2}\sqrt{(k-p)^2}-\frac{S-1}{2}(k-p)+q.$$

Finally, we find

$$\therefore Q=-\frac{S+1}{2}\sqrt{\left(k-\frac{S}{S+1}\right)^2}-\frac{S-1}{2}\left(k-\frac{S}{S+1}\right)+\frac{S}{S+1}. \tag{3.24}$$

Evidently, Eq. (3.24) reads as Eq. (3.13).

(QED)

3.1.2.6 Nagel – Schreckenberg Model (NaSch Model; NS Model)[16]

Although the discussion so far has assumed a deterministic system, real traffic flows are usually influenced by stochastic processes resulting from a driver's decision and the environment surrounding the vehicle. Hence, a plausible CA model should consider stochastic processes. This situation can be considered by a CA model

[16]Nagel and Chreckenberg (1992).

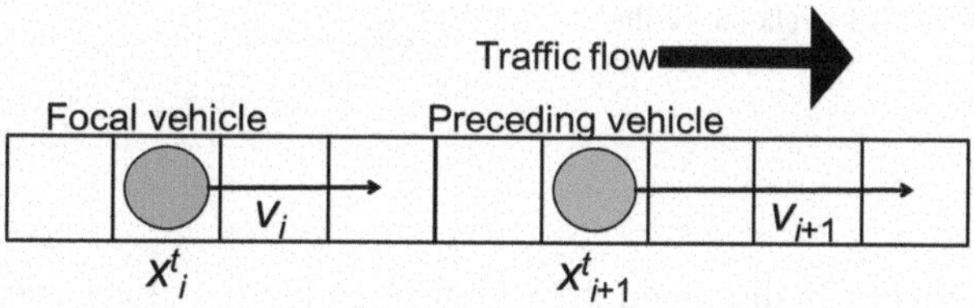

Fig. 3.13 Scheme of the Nagel – Schreckenberg model

because it is based on a set of local rules arising from microscopic concepts and assumes a simulation approach (except for several specific abovementioned models that offer exact solutions); thus, CA models are more flexible than macroscopic or car-following ones (plus specific CA models such as Rule-184, ASEP, ZRP and so forth) premising analytical approach.

The Nagel – Schreckenberg model (NaSch model; NS model) is perhaps the most commonly applied CA model; it considers a stochastic process while still remaining simple. Referring to Fig. 3.13, we describe the model as below:

Rule 1 "Acceleration"

$$v_i^{(1)} = \min\left[V_{\max}, v_i^{(0)} + 1\right]; \tag{3.25a}$$

Rule 2 "Avoid collision"

$$v_i^{(2)} = \min\left[v_i^{(1)}, x_{i+1}^t - x_i^t - 1\right]; \tag{3.25b}$$

Rule 3 "Random brake (RB)"

$$v_i^{(3)} = \max\left[0, v_i^{(2)} - 1\right]; \tag{3.25c}$$

(only if rand() <1 – p is Rule 3 applied, where p is RB provability).

Rule 4 "Moving forward"

$$x_i^{t+1} = x_i^t + v_i^{(3)}. \tag{3.25d}$$

3.1.2.7 Revised S-NFS Model[17]

Although NaSch has been widely accepted as a standard CA model in various simulation studies, it seems too simple to reproduce the rich complexity of real traffic. One state-of-the-art model is the stochastic Nishinari–Fukui–Schadschneider (Revised S-NFS) model,[18] which accounts for motions that are commonly observed in real vehicles: slow-to-start (S2S), quick start (QS), and random braking (RB). As

---

[17]Kokubo et al. (2011).

[18]Sakai et al. (2006).

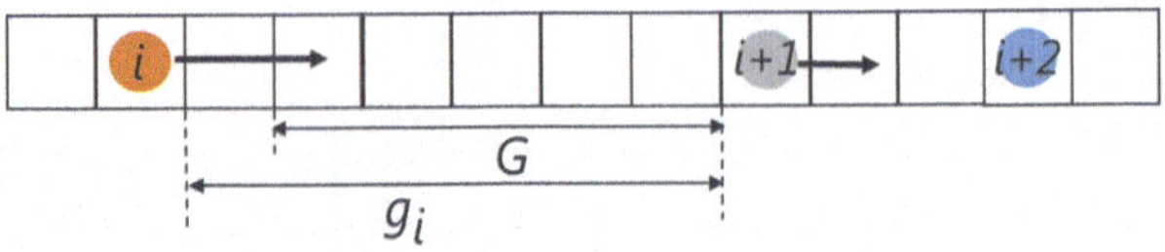

Fig. 3.14 Scheme of the revised S-NFS model

previously discussed, S2S implies an inertial effect, which notably produces metastable states in the fundamental diagrams. Metastable phases are observed and thought to be important features in real traffic flows, where a group of vehicles with less heading distance (implying a reasonable middle traffic density) drive at reasonably high speed (referred to as "platoon driving"), showing maximum flux; however, this is a volatile situation and it can easily and irreversibly phase-shift to a congested phase. QS is the result of acceleration or deceleration by a driver who is anticipating the intentions of both the immediately preceding vehicle and several further ones.

One important issue to note is that variants of the NaSch model, including the S-NSF model, suffer when a high-speed vehicle shows an unrealistic deceleration like a sudden stop by bumping a vehicle ahead of it. This is because NaSch-based models always accelerate regardless of heading distance and do not decelerate gradually when approaching a vehicle ahead, which seems unrealistic. On the other hand, the OV and SOV models offer good performance, because they modulate acceleration or deceleration according to a heading distance related to the traffic density around the focal vehicle. The lack of this particular effect causes an unlikely transition directly from the free-flow (F) to jam phases (J), whereas in reality, this transition generally passes through a synchronous phase (S), as claimed by Kerner in the three-phase theory.[4]

To close this discussion, we introduce the revised S-NFS model as below. This model refines the RB in the S-NFS model to improve the reproducibility of Kerner's three-phase theory. Let us refer to Fig. 3.14 as nomenclature.

Rule 1 "Acceleration"

$$v_i^{(1)} = \min\left[V_{\max}, v_i^{(0)} + 1\right] \tag{3.25e}$$

(only if $g_i \geq G \wedge \ v_i^{(0)} \leq v_{i+1}^{(0)}$; then Rule 1 is applied).

Rule 2 "Slow-to-start (S2S)"

$$v_i^{(2)} = \min\left[v_i^{(1)}, x_{i+s_i}^{t-1} - x_i^{t-1} - s_i\right] \tag{3.25f}$$

(only if rand() $\leq q$; then Rule 2 is applied) and (if rand() $\leq r$, then $s_i = S$; else $s_i = 1$).

Rule 3 "Perspective (Quick start; QS)"

$$v_i^{(3)} = \min\left[{v_i}^{(2)}, x_{i+s_i}^t - x_i^t - s_i\right]. \tag{3.25g}$$

Rule 4 "Random brake (RB)"

$$v_i^{(4)} = \max\left[1, {v_i}^{(3)} - 1\right]. \tag{3.25h}$$

(only if rand() <1 − p_i; then Rule 4 is applied).

$$if(g_i \geq G)$$

$$p_i = P_1 \tag{3.25i}$$

$$if(g_i < G)$$

$$p_i = P_2 \text{ for } v_i^{(0)} < v_{i+1}^{(0)}, \tag{3.25j}$$

$$p_i = P_3 \text{ for } v_i^{(0)} = v_{i+1}^{(0)}, \tag{3.25k}$$

$$p_i = P_4 \text{ for } v_i^{(0)} > v_{i+1}^{(0)}. \tag{3.25l}$$

Rule 5 "Avoid collisions"

$$v_i^{(5)} = \min\left[{v_i}^{(4)}, x_{i+1}^t - x_i^t - 1 + {v_{i+1}}^{(4)}\right]. \tag{3.25m}$$

Rule 6 "Moving forward"

$$x_i^{t+1} = x_i^t + v_i^{(5)}, \tag{3.25n}$$

where x_i^t is the position of vehicle i at time t, v_i is the velocity of vehicle i, $v_i^{(0)}$ is the velocity $v_i^{(5)}$ at the previous time step $t-1$, defined by $x_i^t - x_i^{t-1}$, s_i is the number of preceding vehicles from the ith driver's perspective, g_i is the gap between vehicle i and vehicle $i+1$ (thus, $g_i = x_{i+1}^t - x_i^t - 1$), and V_{max} is the maximum velocity. The notation rand() represents a random number drawn from the uniform distribution on [0, 1]. The quantities G, q, r, S, P_1, P_2, P_3, and P_4 are model parameters. The probability of random braking is given by 1–p_i. We assume that $P_1 > P_2 > P_3 > P_4$. As the standard parameter setting, we assume $q = 0.99$, $r = 0.99$, $S = 2$, $P_1 = 0.999$, $P_2 = 0.99$, $P_3 = 0.98$, $P_4 = 0.01$, $G = 15$, and $k = 3$.

In the rest of this chapter, we apply the revised S-NFS model as our standard framework, since, as mentioned, its plausibility for reproducing realistic driver's behaviors and the Kerner's three-phase theory has been fully validated.[17]

3.1.3 Modeling Lane-Change and Its Validation[19]

To ensure realistic traffic flow, modeling lane-changes is as important as modeling forward driving.

Incidentally, sub-models accounting for a vehicle's lane change on the basis of the CA framework and several alternatives have been presented. All these alternatives basically comprise two conditions: incentive and security criteria. The incentive criterion refers to the driver's motivation for changing lanes, while the security criterion refers to a threshold condition of whether a lane change is safe. Both criteria are expressed by inequalities defined by several specified distances between a focal vehicle and its preceding vehicles in both the current and destination lanes or the following vehicle in the destination lane.[20] Regarding the incentive criterion, most previous studies only refer to the distances between vehicles in both the current and destination lanes. Kerner and Klenov,[21] however, developed a lane-changing CA model based on the velocity difference between neighboring vehicles. Moreover, we have developed a model that defines the incentive criterion considering both the distance and velocity difference between the preceding vehicles in the current and destination lanes.[22] All the aforementioned sub-models define each of the incentive criteria empirically. Although they have been expected to work plausibly, they have not been validated based on behaviors of real drivers. In particular, the answer to the question of whether a driver is motivated to make a lane change because of insufficient distance in the current lane but a sufficient one in the destination lane (only distances motivated) or because of a negative-velocity difference in the current lane but a positive one in the destination lane (meaning the driver can accelerate in the neighboring lane; velocity-difference motivated) is not clear, owing to a lack of field-measurement data.

3.1.3.1 Measurement Setup and Data Handling

Our field measurements were conducted between May 3–6, 2014, by looking over a section of the Kyushu Expressway using several cameras connected independently to notebook-type PCs for data acquisition. As shown in Fig. 3.15a, the observation site was positioned at the clifftop of Mt. Otogana, where the cameras could appropriately view all four lanes of traffic. However, we focused on the two lanes with traffic flow in the south direction. As shown Fig. 3.16b, an 80-m-long control volume was placed on the expressway and a cross section of the expressway was selected to count the number of passing vehicles.

[19]Fukuda et al. (2016).

[20]Many previous works have examined this question, including: Gundaliya et al. (2008), Tanaka et al. (2008), Lan et al. (2009).

[21]Kerner and Klenov (2009).

[22]Kukida et al. (2011).

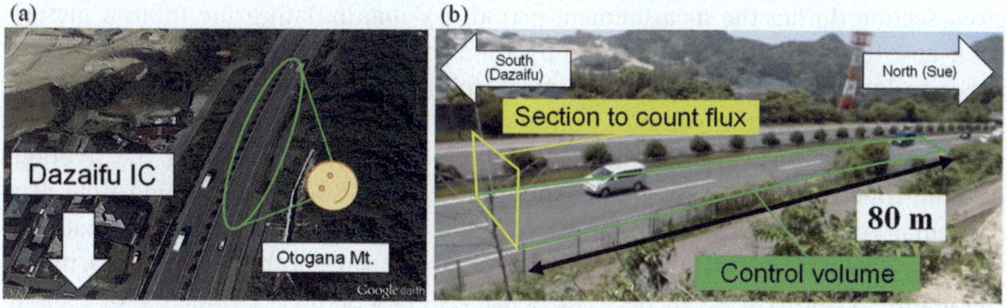

Fig. 3.15 The left panel illustrates the measurement site on Kyushu Expressway, where cameras were installed on the clifftop of Mt. Otogana, 2.1 km north of the Dazaifu Interchange (Otogana, Onojo City, Fukuoka Prefecture). The image is taken from Google Earth™. The right panel shows the definitions of the control volume (CV) and the cross section used to count the number of passing vehicles for the measurements of traffic flux, average velocity, density, lane usage ratio, and frequency of lane change. The length of the CV is 80 m

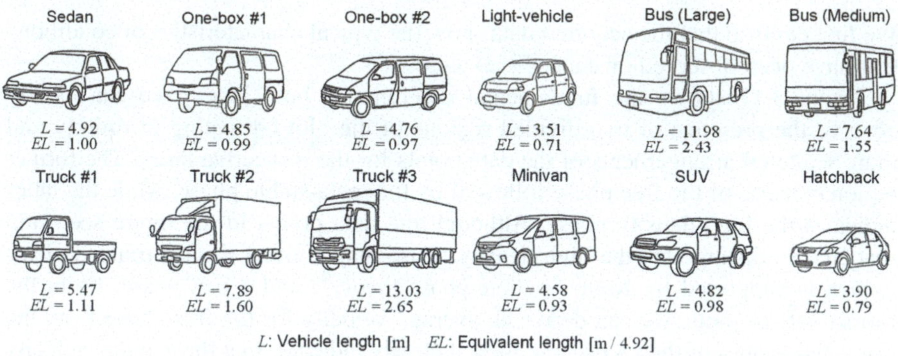

Fig. 3.16 Actual vehicle lengths and equivalent lengths (*EL*s) for each of the 12 vehicle classes

We follow the method of our previous work[3] to handle the acquired data to obtain traffic flux, density, and lane-usage ratio.

Unlike loop-detector data, motion-picture data allow analysis of each vehicle via the Lagrangian approach.

We first define the 12 vehicle classes illustrated in Fig. 3.16 and their respective equivalent lengths (*EL*s), normalized to the actual length of a standard sedan-type vehicle, i.e., $L_{sedan} = 4.92$ m.

Next, we separately count the number of vehicles belonging to each of the respective classes passing through the cross-section set at the downward of the control volume per unit time (see Fig. 3.15b). Specifically, we capture $\Sigma N_i'$ [(vehicles)/(15 s)], where i indicates one of the twelve classes. In the present study, we rely on the total equivalent number of vehicles, $\Sigma N_i' * EL_i$ $(=\Sigma N_i)$ [(vehicles)/(15 s)], instead of $\Sigma N_i'$, where, in a nutshell, $\Sigma N_i'$ is converted into the total number of standard sedan-type cars based on their lengths. ΣN_i can be transformed into the total flux $q = \Sigma q_i$ [(vehicles)/s]. Simultaneously, we focus on each of the vehicles passing through the cross section during the measurement period. By manipulating the motion picture back and forth, we derive the running time of a particular vehicle between the upper to lower boundaries of the control volume and transform this information into the velocity of the vehicle. Following this data analysis, we can obtain the average velocity by dividing the control volume, and by finding the average running time of vehicles over a 15-speriod, i.e., $v = 80/(\Sigma$ the running time of a particular vehicle over a 15-s period/$\Sigma N_i')$ m/s. Consequently, we can estimate the density ρ [m/m] (=[ND]) over a 15-s period to measure both flux and velocity by calculating ρ [m/m] $= L_{sedan} \cdot q/v$. The density, ρ, is separately evaluated for each of the two lanes.

The lane-usage ratio is calculated by taking the arithmetic mean of 4 successive time-sections of density over 4* 15-[s] (= 60-[s]).

3.1.3.2 Basic Characteristics of the Flow Field

We first confirm that the acquired data show the typical characteristics of conditions that have been observed on Japanese expressways.

Figure 3.17 shows the fundamental diagram for both lanes. We can clearly observe the presence of two divided regions of the plot consisting of orderly and more scattered arrangements of the data points for the respective lanes. The former region consists of the free phase followed by the meta-stable phase, while the latter indicates the high-density phase. Although the data points in the figure seem too scattered to distinguish, the high-density phase may contain a synchronized phase (which is supported by Kerner's three-phase theory[4]) and a jam phase. From the former sets of plots, we can derive an average velocity for the free phase from the respective slopes of the regression lines. It clearly indicates that the average velocity in the slow lane is lower than that in the fast lane. The corresponding two slopes are approximately comparable to the maximum velocity of 100 km/h on a Japanese expressway.

As shown in Fig. 3.17, the peak flux of the fast lane is obviously larger than that of the slow lane. This implies that the fast lane allows vehicles to travel at a higher speed in a compact manner compared with the slow lane; this is known as a "platoon formation."

The fundamental diagram that we obtained in the present study shows analogous features to those of previous studies of conditions on other Japanese expressways[3,23]

Figure 3.18 shows the relationship between the lane-usage ratio and mean density of the two lanes expressed in physical dimensions [(vehicles)/km/lane] and the mean normal density as a secondary horizontal axis. A lager ratio indicates that a majority of vehicles travel in the slow lane. What we can note as a general tendency is that a majority of vehicles stay in the slow lane as long as the mean density remains

[23]Tadaki et al. (2013).

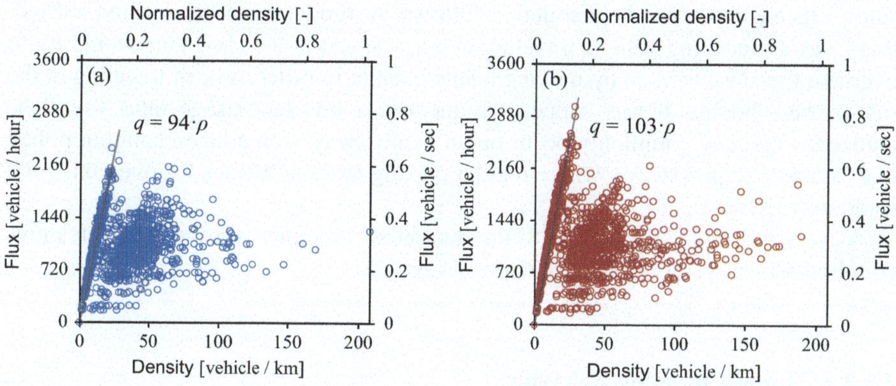

Fig. 3.17 Fundamental diagrams showing the traffic-flux density; (**a**) for the slow lane and (**b**) for the fast lane. Here, it should be noted that there is an anomalous data plot whose normalized density is greater than 1 (approximately 1.02) in a jam phase in 3.17a. These erratic data come from minute measurement errors of both the velocity and flux

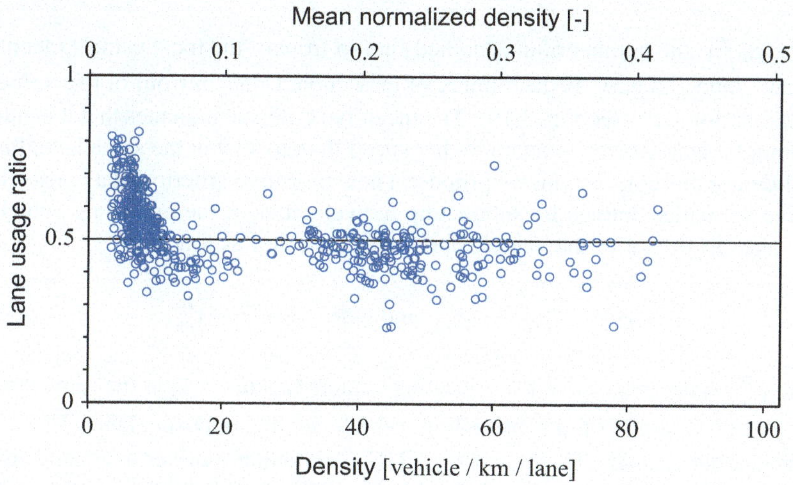

Fig. 3.18 Lane-usage ratio vs mean density of the two lanes with mean normal density as a secondary horizontal axis

sufficiently low; and even with increasing density, there seems no obvious difference in the observed trend despite a larger scattering between the slow and fast lanes. This observation is justified because there is no incentive to change lanes from the slow lane to the fast one as long as every vehicle travels at maximum velocity. The flow field at a relatively high density does not allow frequent lane changes; this reduces the advantage of staying in the fast lane. One notable point is that there is a density range between the low and high densities where the majority is in the fast lane rather than in the slow one. This "inversion" tendency was also reported in our previous study[3]. Its existence can be justified as follows. At moderate traffic density, a driver that faces a preceding vehicle traveling at a low speed in the slow lane would try to overtake the slow vehicle by making a lane change in order to keep traveling at the free-phase velocity. If this happens frequently, a low lane-usage ratio would be realized. This may commonly occur on an expressway with a lane-changing policy like Japanese expressways, where regular driving lanes and lanes for overtaking are rigorously defined.

As seen in Figs. 3.17 and 3.18, the dataset we obtained shows typical features commonly observed on Japanese expressways.

3.1.3.3 Lane-Changing Behavior

Here, the incentive criterion in a typical conventional lane-changing sub-model of the CA model is described[20,22], with

$$gap_p^f \leq v_i^{(p)} \text{ and } gap_n^f \geq v_i^{(p)}, \tag{3.26}$$

where gap_p^f is the number of unoccupied sites in front of the focal vehicle (agent i) in the same lane and gap_n^f is the number of unoccupied sites in front of this vehicle in the destination lane (see Fig. 3.19). The incentive criterion indicates that it is rational to attempt a lane change to attain higher speed than to stay in the same lane. Instead of following this convention, we propose a new incentive criterion that considers the relative velocities defined by subtracting the velocities of the preceding vehicles in both lanes from that of the vehicle under consideration as follows[22]:

$$gap_p^f \leq v_i^{(p)} - v_{i+1}^{(p)} \text{ and } gap_n^f > v_i^{(p)} - v_{i+1}^{(n)}, \tag{3.27}$$

where $v_{i+1}^{(p)}$ is the velocity of the preceding vehicle (agent $i + 1$) in the same lane and $v_{i+1}^{(n)}$ is the velocity of the preceding vehicle in the opposite lane. The crucial difference between Eq. (3.26) and Eq. (3.27) is whether a driver evaluates his own velocity or the relative velocity defined by the difference between his own velocity and that of the preceding vehicle. Kukida et al.[22] shows the results of numerical

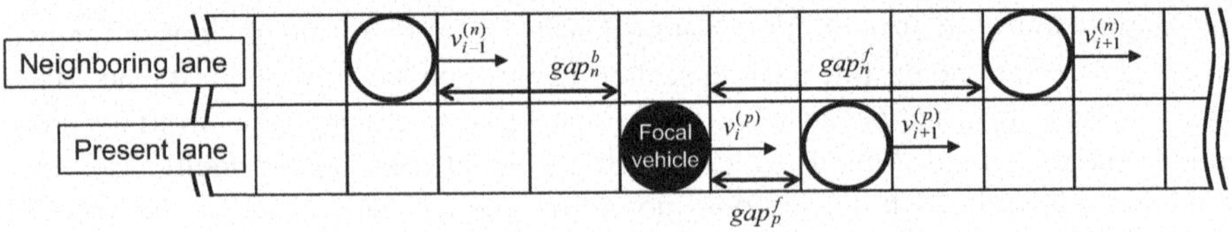

Fig. 3.19 The assumed situation of a car changing lanes, showing the definitions of gaps and velocities. The black circle indicates the focal vehicle and the white circles indicate neighboring vehicles

simulations and concluded that the latter sub-model, i.e., Eq. (3.27), results in more frequent lane changes, especially in the high-density region, than does Eq. (3.26), because a low velocity hardly agrees with Eq. (3.26) but possibly satisfies Eq. (3.27). Real-life experience suggests that there might be several aggressive drivers who seek to overtake preceding cars through forceful lane changing, even in a traffic-congestion situation. Qualitatively speaking, we can apply either Eq. (3.26) or Eq. (3.27) depending on various situations of traffic flows, but there is a lack of proven knowledge regarding this point. In this study, we analyze the measurement survey of real traffic flow in order to provide an incentive criterion for lane-changing events from the perspective of the gap (that is, distance) and the (relative) velocity with respect to the preceding vehicles motivated by their works[22].

Figure 3.20 shows the relationship of the gaps between the focal vehicle and a preceding vehicle in the present and neighboring lanes, i.e., gap_p^f and gap_n^f, respectively. Panel (a) shows the case for a lane change from the slow lane to the fast one and panel (b) shows that from the fast lane to the slow one. Panel (c) shows the schematic, which describes how to calculate each vehicle's velocity and to obtain gap_p^f and gap_n^f.

The data consist of all actually observed lane changes. Thus, we cannot use any arguments regarding the lane-change frequency or probability; i.e., how frequently a lane change takes place under certain gap conditions. But, again, the data shown here indicate the conditions for which a lane change actually takes place.

The shown data do not contain any information about the vehicle's velocity. It is unnecessary to discuss whether all data correspond to the incentive criterion because it would mean that Eq. (3.26) is valid irrespective of the value of $v_i^{(p)}$, as long as a data point is observed above the 45° line.

One crucially important observation in Fig. 3.20 is that the majority of actual lane-change events are plotted above the 45° line in the case of a vehicle changing from slow to fast lanes. More precisely, this tendency seems clear for the data outside the green-highlighted region, which implies that a vehicle can maintain sufficiently large distances from preceding vehicles in both lanes. However, as long as a vehicle maintains an insufficient distance, the incentive criterion expressed by Eq. (3.26) does not seem valid. This point will be addressed below in the context of Fig. 3.21.

Figure 3.20b clearly shows that the incentive criterion of Eq. (3.26) does not come into effect. Obviously, actual lane-change events take place rather randomly and irrespective of gap_p^f and gap_n^f. That seems natural because Japanese traffic regulations determine the difference between a slow and fast lane and a vehicle must principally travel in the slow lane and is allowed to use the fast lane only to overtake a slow preceding vehicle in the same lane. Thus, a vehicle is expected to return to the usual running lane, i.e., the slow lane, immediately after completing an overtaking event. Therefore, the incentive criterion cannot be applied to a case when a vehicle changes from a fast lane to a slow lane as long as the Japanese expressway policy is considered.

Figure 3.20a implies that Eq. (3.26) does not apply if a vehicle has less than a sufficient distance from the preceding vehicles in the present and neighboring lanes.

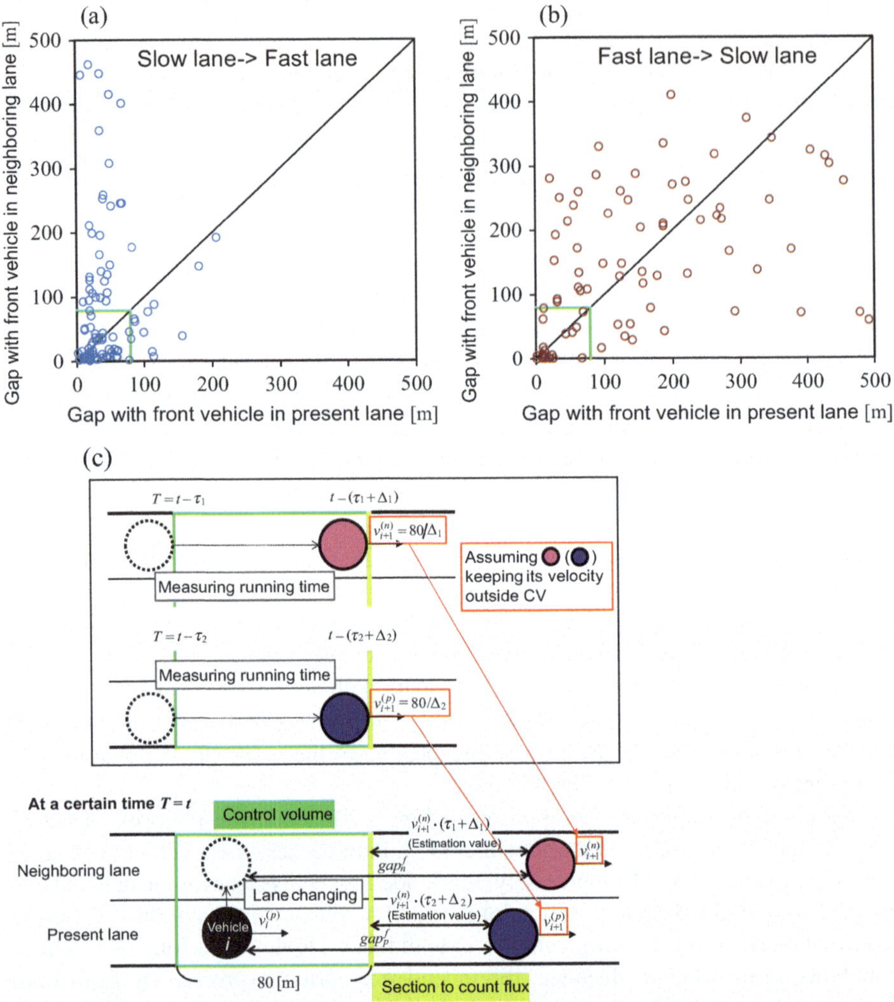

Fig. 3.20 The gaps between the focal vehicle and neighboring vehicles in the present and neighboring lanes for lane change from (**a**) the slow lane to the fast one and (**b**) the fast lane to the slow one. The green-highlighted regions indicate the control volume. Note that the gaps are estimated by assuming that the foregoing vehicles in each lane keep their own velocities, even if they go outside the control volume, which were measured when they passed inside the control volume shown in (**c**)

Figure 3.21 shows an enlarged image of the green-highlighted region of Fig. 3.20a, where both distances are less than the length of the control volume (80 m); the plot is classified into three classes. A red plot means a vehicle for which $v_{i+1}^{(p)} - v_i^{(p)} < 0$ and $v_{i+1}^{(n)} - v_i^{(p)} > 0$; this implies that the vehicle unfailingly accelerates during a

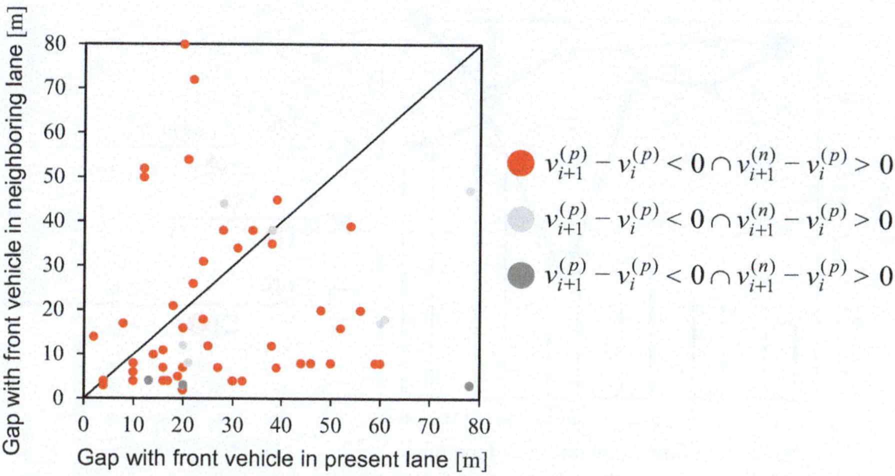

Fig. 3.21 The gaps between the focal vehicle and neighboring vehicles in the present and neighboring lanes when changing from fast to slow lanes. The figure shows the enlarged green-highlighted region of Fig. 3.20a. Red datapoints indicate vehicles with $v_{i+1}^{(p)} - v_i^{(p)} < 0$ and $v_{i+1}^{(n)} - v_i^{(p)} > 0$; this implies that the vehicle unfailingly accelerates under lane changes. Light-gray datapoints indicate vehicles with $v_{i+1}^{(p)} - v_i^{(p)} < 0$ and $v_{i+1}^{(n)} - v_i^{(p)} < 0$; this implies the vehicle changes lanes but has no possibility of acceleration. Dark-gray datapoints indicate vehicles with $v_{i+1}^{(p)} - v_i^{(p)} > 0$ and $v_{i+1}^{(n)} - v_i^{(p)} > 0$; this implies that the vehicle changes lanes but can accelerate in the present lane as well. No data are classified as $v_{i+1}^{(p)} - v_i^{(p)} > 0$ and $v_{i+1}^{(n)} - v_i^{(p)} < 0$

lane change. A light-gray plot indicates a vehicle for which $v_{i+1}^{(p)} - v_i^{(p)} < 0$ and $v_{i+1}^{(n)} - v_i^{(p)} < 0$; this implies that there is no possibility of acceleration during a lane change. A dark-gray plot means $v_{i+1}^{(p)} - v_i^{(p)} > 0$ and $v_{i+1}^{(n)} - v_i^{(p)} > 0$; this implies that the vehicle does change lanes but can accelerate while staying in the present lane. None of the data are classified by the conditions $v_{i+1}^{(p)} - v_i^{(p)} > 0$ and $v_{i+1}^{(n)} - v_i^{(p)} < 0$.

At a glance, we note that a vast majority of the acquired data fall into the class with $v_{i+1}^{(p)} - v_i^{(p)} < 0$ and $v_{i+1}^{(n)} - v_i^{(p)} > 0$. Figure 3.22 shows three fractions as a function of the gap range with a preceding vehicle in the present lane when changing from the slow lane to the fast one. A diamond-shaped datapoint indicates the proportion of vehicles with $v_{i+1}^{(p)} < v_{i+1}^{(n)}$ to the total number of vehicles in each gap range—i.e., the fraction of vehicles that change lanes motivated by the velocities of the preceding vehicles. The cross datapoint indicates the proportion of vehicles with $v_{i+1}^{(p)} - v_i^{(p)} < 0$ and $v_{i+1}^{(n)} - v_i^{(p)} > 0$ (the total number of the red datapoints shown in Fig. 3.21) to the total number of vehicles in each gap range with preceding vehicles in their present lane; this is the fraction of the vehicles which change from the slow lane to the fast one motivated by the relative velocities of preceding vehicles. A bar

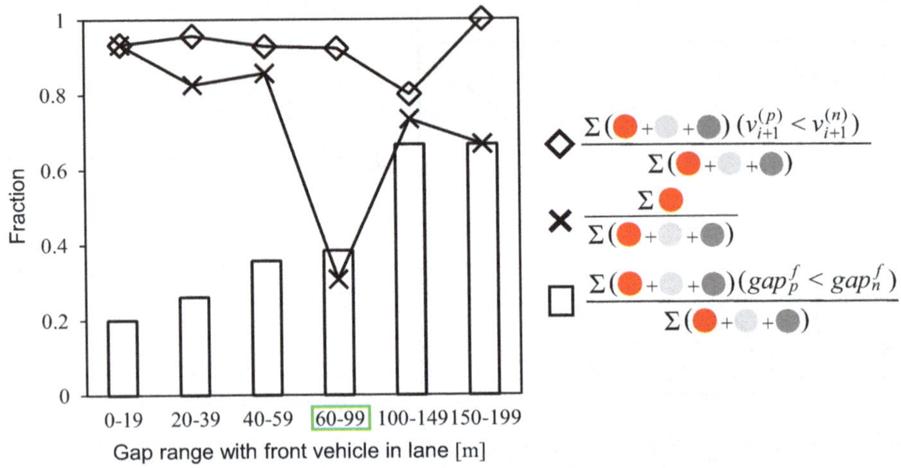

Fig. 3.22 Three fractions as functions of the gap range with the preceding vehicle in the present lane when changing from slow to fast lanes. Red, light-gray, and dark-gray circles in the legends represent the same meaning as in Fig. 3.21. A diamond-shaped datapoint indicates the proportion of vehicles with $v_{i+1}^{(p)} < v_{i+1}^{(n)}$ to the total number of vehicles in each gap range; this is the fraction of vehicles that change lanes motivated by the velocities of the preceding vehicles. The cross datapoint indicates the proportion of vehicles with $v_{i+1}^{(p)} - v_{i}^{(p)} < 0$ and $v_{i+1}^{(n)} - v_{i}^{(p)} > 0$ (the total number of red datapoints) to the total number of vehicles in each gap range—i.e., the fraction of vehicles that change the lane motivated by the relative velocities with respect to the preceding vehicles. A bar indicates the proportion of vehicles in the region of above the 45° line in Fig. 3.20 (the total number of datapoints with $gap_p^f < gap_n^f$) to the total number of vehicles in each gap range with preceding vehicles in their lane—i.e., the fraction of vehicles that change lanes motivated by gaps with respect to the preceding vehicles. The green-highlighted range corresponds to the critical gap, at which the incentive criterion changes from referring to the velocities and(/or) the relative velocities with respect to the preceding vehicles, to taking into account the distances from preceding vehicles in addition to the above two factors

chart indicates the proportion of the total number of vehicles in the region above the 45° line in Fig. 3.20 (the total number of data points with $gap_p^f < gap_n^f$) to the total number of vehicles in each gap range with a preceding vehicle in their present lane; this is the fraction of vehicles that change from the slow lane to the fast one motivated by gaps with respect to preceding vehicles.

As can be seen from Fig. 3.22, it turns out that both the relative velocities and the velocities of preceding vehicles themselves are building blocks for the incentive criteria for the entire gap range. When the gap exceeds 60–99 m (highlighted in green), the fraction with regard to the distances illustrated by bars becomes greater than 0.5, such that the distance from the preceding vehicles is also included in the incentive criterion, in addition to the two factors mentioned above. Though the behaviors of both the cross- and diamond-shaped datapoints in the Fig. 3.22 yield important information for establishing the incentive criterion, they do not allow us to identify the critical gap [m] at which the incentive criterion changes (i.e., the boundary of the phase transition). Alternatively, we clearly see that the bars in the Fig. 3.22 become saturated as the gap range exceeds 60–99 m, leading us to conclude that the gap range of 60–99 m is critical.

To summarize, we found 3 factors composing the incentive criterion; (I) the velocities of preceding vehicles, (II) the relative velocities with respect to preceding vehicles, and (III) the distances (gaps) with respect to preceding vehicles. In addition, we identify the phase of the incentive criterion; for the region of short-gap range (smaller than the critical gap, 60–99 m), the combination of (I) and (/or) (II) constitute one phase, whereas for the region of long-gap range greater than the critical gap, factor (III) is added to (I) and (II) and all of these constitute another phase. Here, the constituents of the incentive criterion are expressed as follows:

$$\text{(I)}\ \ v_{i+1}^{(p)} < v_{i+1}^{(n)}, \tag{3.28}$$

$$\text{(II)}\ \ 0 < v_i^{(p)} - v_{i+1}^{(p)} \ \text{ and }\ 0 > v_i^{(p)} - v_{i+1}^{(n)}, \tag{3.29}$$

$$\text{(III)}\ \ gap_p^f < gap_n^f. \tag{3.30}$$

In this connection, Eq. (3.29) is identical to the condition for the red datapoint in Fig. 3.21, and also identical to Eq. (3.27) with its left-hand side replaced by 0.

3.1.3.4 Summary

To determine what incentive criterion triggers a lane-change event, we collected actual traffic-flow data for a Japanese expressway. These data were used to confirm the fundamental diagrams and lane-usage ratios to show typical features for Japanese expressways.

A change from a fast lane to a slow lane occurs regardless of the incentive criterion because Japanese traffic regulations demand that a vehicle must always travel in the slow lane except during the transient time when they are trying to overtake the preceding vehicle.

However, there two are incentive criteria for changing from the slow lane to fast lane underlying the actual events—that is, we can identify the phase of the incentive criterion. For the region of short gap range (smaller than the critical gap, 60–99 [m]), the combination of (I) the velocities of the preceding vehicles and (II) the relative velocities with respect to the preceding vehicles constitute one of phases; for gap ranges greater than the critical value, factor (III), the distances from the preceding vehicles, is added to (I) and (II), constituting another phase.

In addition, we found that the results of this study justify the incentive criterion previously assumed for many CA models, e.g., that of Kukida et al.[22] This comes into effect in actual events, but should be switched depending on the distances from the preceding vehicles. In this study, however, we do not directly validate Eqs. (3.26) or (3.27) during the analysis of our measurement survey, but we can nevertheless conclude that Eq. (3.29) becomes a special case of Eq. (3.27), replacing gap_p^f and gap_n^f by 0. This is because, as is the case for most of CA models including their works, both the vehicle velocity and *gap* at one time-step are treated as having the same dimension. Obviously, these variables have different dimensions in real traffic-flow systems.

Finally, we comment briefly upon the dramatic decrease of the cross datapoint at 60–99 [m] in Fig. 3.22. We are unable to find any rationale for this anomaly; hence we should hold a thorough investigation into the causes of such strange behavior in future work, for example, by performing additional observations survey deeper analysis of lane-changing behavior in two lanes on an expressways in a similar situation.

3.2 Social Dilemma in the Rote-Selection Problem[24]

Again, let me say that social concern regarding traffic has been growing over the last decades. Leading-edge information, communication, and internet technology has pushed the development of Intelligent Transportation Systems (ITS), which aim to maximize the urban traffic efficiency by mitigating traffic jams. A traffic flow results from collective phenomena from numerous vehicles; however, maximizing each individual benefit may not necessarily indicate a maximum social or global benefit. Here, the term "benefit" implies the objective variable for the current discussion, which is a travel time of one single origin & destination (OD) journey. If the traffic flow results from the competition for a finite resource ("road") among many drivers seeking shorter and more comfortable driving times, then it should be determined whether a social dilemma originates from the intentions of drivers. One simple example of such a situation is the so-called route-selection problem, which has been extensively studied[25,26] One study involved modeling route selection motivated by the application of ITS. A driver is exposed to a route-selection problem whereby they must choose between two paths reaching the same destination with the same distance. A driver can use traffic information, which may be provided through either the vehicle-information-and-communication-system (VICS) device or a simple signage board. The traffic information that can increase traffic flux for maximizing social efficiency is important. Here, we investigate whether a social dilemma exists in a particular traffic flow by establishing a CA model; if one does exist, we determine whether it belongs to the PD, Chicken, or other classes.

[24]Tanimoto and Nakamura (2016).

[25]Whale et al. (2000).

[26]Xiang and Xiong (2013).

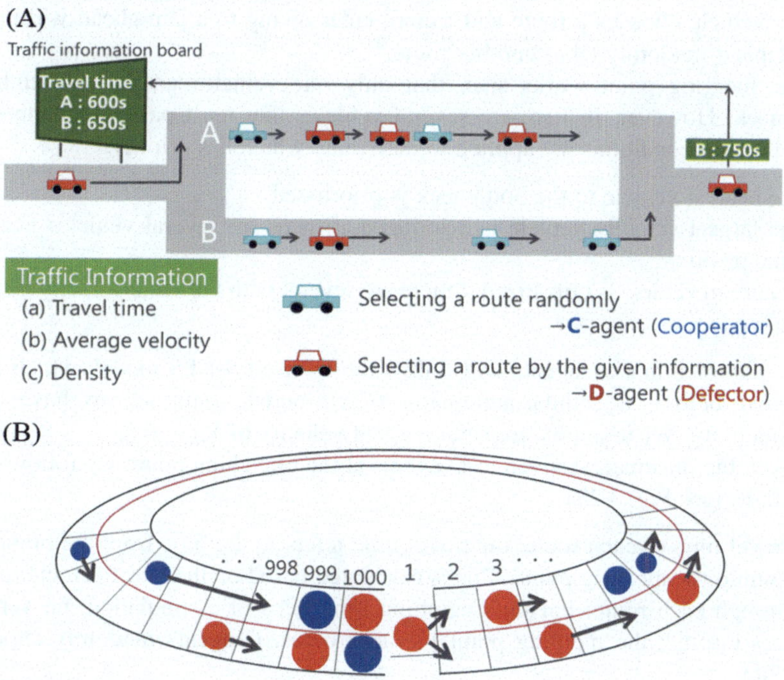

Fig. 3.23 Panel (**a**); Model for route selection whereby cooperators and defectors coexist, with both selecting their next route randomly or by referring to either (a) travel time, (b) average velocity, or (c) density. Panel (**b**) shows the present CA model for panel (**a**), where the red and blue circles indicate defectors and cooperators, respectively. The two lanes separated by the red line mean that vehicles are not allowed to change lanes in cells labeled #2 to #1000

3.2.1 *Model Setup*

Our model is based on that of Xiang.[26] We assume that a traffic system has two routes, A and B,[27] each of the same length $L = 1000$ cells and having a single lane. The routes diverge at the entrance and converge them at the exit, although the terms "entrance" and "exit" are trivial because the system is cyclic, as shown in Fig. 3.23.

At the branching point, a vehicle is given traffic information. We defined two types of agents. One is the defector (D), who always moves fast and takes routes with the least travel time, largest average velocity, and lowest average density. The other is the cooperator (C), who is less interested in moving fast and chooses routes randomly.

[27]Although we have not discussed the case of three alternative routes, we have confirmed that the general conclusion that can be drawn is basically the same as in the case of two routes, as reported herein.

If a vehicle chooses a route and cannot enter owing to a jam ahead within five timesteps, it randomly takes another route.[28]

The merging point works such that only one vehicle can pass through the bottleneck. However, if there are several vehicles that must pass simultaneously, the following conditions are applied to determine which has priority.

1. The nearest vehicle to the bottleneck is prioritized.
2. The largest-velocity vehicle is prioritized if there are several vehicles near the same position.
3. A random vehicle is prioritized if there are several with the same velocity near the same position.

We assumed that all agents will follow the Revised S-NFS model, which is, as discussed before, a cellular-automaton (CA) model confirmed to have good reproductivity. We assume a maximum agent velocity of $V_{\max} = 5$.

Given the information given to D agents at the branching point, we follow three conditions (see Fig. 3.23):

(a) Travel time of each route: the travel time refers to the time from the branching point to the merging point. This is calculated based on the latest vehicle to pass through each route. For the beginning time-steps of a simulation, no vehicles pass through the merging point via any routes. D agents randomly choose a route.
(b) Average velocity of each route: for each route, the velocity is averaged over all vehicles passing through each route.
(c) Density of each route: the density is determined on the basis of the number of vehicles passing through each route.

As previously mentioned, we use the cyclic boundary condition to maintain constant vehicle density during a single simulation run. Each simulation run repeats steps (i)–(iii), with steps (ii)–(iii) correspond to one time step.

(i) N_S vehicles are generated and placed at random positions in the system. The C (D) agent fraction among N_S is P_c $(1 - P_c)$.
(ii) The next-step velocity of all agents is determined using the revised S-NFS model.
(iii) All agents update their positions in the system.

In numerical experiments, we set the system length to $L = 1000$ and assume the following values for model parameters: $q = 0.99$, $S = 2$, $P_1 = 0.999$, $P_2 = 0.99$, $P_3 = 0.98$, $P_4 = 0.01$, $G = 15$, $k = 3$, and $P_{LC} = 1$. We vary N_S from 50 to 950 in increments of 10 vehicles, which control the traffic density in system ρ. We also

[28] Although not reported, we can extend the model for three or more alternative routes as mentioned in the text. In those cases, this part of our model was modified as in the text. If a defective vehicle chooses one route, but cannot enter it owing to a jam within five time-steps ahead, it instead takes the second-best route. Likewise, if a cooperative vehicle chooses one route but cannot enter it due to a jam ahead with five time-steps, it randomly takes one of the remaining alternative routes.

varied P_c from 0 to 1 in increments of 0.1. We varied r, one of the model parameters in the revised S-NFS model, which controls the quick-start (QS) effect. As the presumed value of r becomes larger, more vehicles have longer perspectives ($s_i = S = 2$) than the default setting ($s_i = 1$). This causes smooth acceleration and deceleration, resulting in a large traffic flux due to the low possibility of a stop-and-go wave.

All results were determined from 100 independent realizations. We determined the average velocity of each agent as L/(travel time),and averaged those over all C agents (D agents) to find the average payoff of each strategy. For the social payoff, we use the time-averaged traffic flux.

3.2.2 Results and Discussion

Figure 3.24 shows the payoff-structure functions when travel time is provided as traffic information, wherein the cooperator's average payoff (average velocity of C; blue triangle), defector's average payoff (average velocity of D; red circle), and social average payoff are expressed by an average flux (green diamond). In panel (a) $r = 0.1$ and $\rho = 0.2$; (b) $r = 0.1$ and $\rho = 0.4$; (c) $r = 0.7$ and $\rho = 0.2$; and (d) $r = 0.1$ and $\rho = 0.03$. The crossing points of the cooperator's and defector's payoff functions (open circles) indicate the Nash equilibrium (NE), while the open squares represent the social-maximum payoff (SM). In evolutionary game theory as we discussed in Chap. 2, we can evaluate whether a certain game has a social dilemma by determining if NE is consistent with SM. The game has to contain either a PD, Chicken, or Stag Hunt social dilemma if NE is inconsistent with SM. If an NE appears at $P_c = 0$, the social dynamics will be absorbed by the all-defectors-state; namely, the defection-dominant (D-dominate) state, which corresponds to the PD. If an NE appears at P_c, as shown in Fig. 3.24a, its dynamics would be absorbed by the crossing point, the internal equilibrium (polymorphic equilibrium that belongs to the Chicken game class). If the value of P_c is larger than the crossing point (internal equilibrium), the cooperator's fraction must decrease because the defector's payoff is superior to the cooperator's payoff. On the other hand, if a cooperator's fraction is less than the internal equilibrium, the situation wherein the defector's payoff is always inferior to that of the cooperators, the cooperator's fraction increases. Therefore, the NE that originally implies a point absorbing a social dynamic to the end of an evolutionary path must be an internal equilibrium. If this is the case, it is called a Chicken game. Moreover, Fig. 3.24a should be classified as a multiplayer Chicken (n-Chicken) game. If a crossing point appears halfway to P_c but the cooperator's (defector's) payoff function is superior to that of the defectors (cooperators), in a region of larger (smaller) P_c, this crossing point works as a "source of equilibrium" unlike that in a Chicken game, where the crossing point becomes a "sink of equilibrium". Therefore, its social dynamics would be attracted to either an all-defectors-state ($P_c = 0$) or an all-cooperators-state ($P_c = 1$),

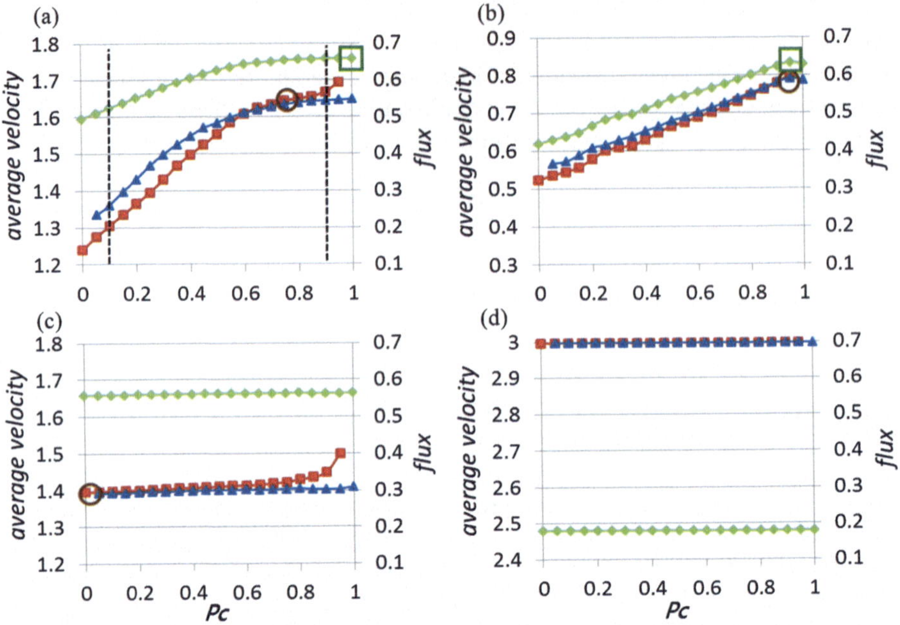

Fig. 3.24 Payoff-structure functions when the travel time is provided as traffic information, showing the cooperator's average payoff (average velocity of C; blue triangle), defector's average payoff (average velocity of D; red circle), and socially averaged payoff average flux (green diamond). We assumed (**a**) $r = 0.1$ and $\rho = 0.2$, (**b**) $r = 0.1$ and $\rho = 0.4$, (**c**) $r = 0.7$ and $\rho = 0.2$, and (**d**) $r = 0.1$ and $\rho = 0.03$. These four conditions are depicted with four crosses in Fig. 3.26a. The crossing point of the cooperator's and defector's payoffs, which is indicated by the open circle, is the Nash Equilibrium (NE), while the open square is the social maximum (SM). Two vertical dotted lines in panel (a) show $P_c = 0.1$ and 0.9, of which spatiotemporal diagrams are shown in Fig. 3.25

depending on the initial fraction of cooperators in an evolutionary path. These dynamics are known as "Bi-stable"—i.e., this is a Stag Hunt game.

Although Fig. 3.24b shows an internal equilibrium (NE) appearing on hallway P_c, we should not regard this as a Chicken game. As NE is almost consistent with SM at $P_c = 0.9$, almost no drivers experience a dilemma. Thus, this game structure should be classified as "polymorphic trivial".

Figure 3.24c indicates the NE appearing in all-defectors states. However, this is not the PD because the social payoff seems insensitive to P_c. Hence, this is a neutral game according to our previous work,[29] wherein we observe no social dilemma due to insensitivity of the social payoff to the cooperation fraction.

In Fig. 3.24d, we observe no sensitivity to social payoff and no difference between the cooperator's and defector's payoffs, which is called a game-less structure.

[29]Tanimoto et al. (2014a).

By confirming each payoff-structure function on various combinations of r and ρ, we confirmed that only the dilemma class is observed to be Chicken. Otherwise, what we observe is either trivial, neutral, or game-less, as shown in Fig. 3.24b–d. Unlike our previous studies[29,30,31,32] we found that the social dilemma is not a PD but a Chicken game. This can be understood as below. Here, the agents compete for a common limited resource (or public good). The common resource implies the total traffic rates of the two routes, which can be maximized if each route absorbs 50% of agents. If the information given to defective agents is misleading (for example, affected by time-delay), a free-rider may select a wrong choice that rather brings more congestion in that particular route, contributing to poor efficiency. This situation is similar to the El Farol Bar Problem,[33] where the most efficient (least congested) situation occurs when each of the two bars shares half of the demand (as long as the two bars have the same capacity), which is SM. If we consider a 2×2 game, where the payoff matrix consists of four elements (payoff of mutual defection (punishment; P), payoff of one successfully exploiting another (temptation; T), payoff of one being exploited by another (sucker; S), and payoff of mutual cooperation (reward; R)), the scenario mentioned above can be described with the Chicken-type dilemma, where the conditions $D_g = T - R > 0$, $D_r = P - S < 0$, and $S + T > 2R$ are all met,[34,35] The key point is that "$S + T > 2R$" assures that the coexistence of 50% cooperators and 50% defectors is preferred over mutual cooperation (100% cooperators) and mutual defection (100% defectors). ST-reciprocity is superior to R-reciprocity in these game structures,[36,37] (see footnote about ST-reciprocity and R-reciprocity[38]).

Figure 3.25 shows the spatiotemporal diagrams of routes A and B for $P_c = 0.1$ and 0.9, assuming that $r = 0.1$ and $\rho = 0.2$, which was confirmed by the Chicken dilemma, as shown in Fig. 3.24a. Panel (a) shows that the less-cooperative case ($P_c = 0.1$) contains intermittent congestion, where a heavy traffic jam takes place caused by a huge stop-and-go wave. We can confirm that when route A was less congested, a traffic jam occurred in route B, implying that a large traffic jam can alternately occur in either route A or B. The traffic jam is triggered by the

---

[30] Yamauchi et al. (2009).

[31] Nakata et al. (2010).

[32] Tanimoto et al. (2014b).

[33] Arthur (1994).

[34] Tanimoto and Sagara (2007a).

[35] Wang et al. (2015).

[36] Tanimoto and Sagara (2007b).

[37] Wakiyama and Tanimoto (2011).

[38] A set of Chicken-type games satisfying $T + S > 2R$, such as Leader and Hero games, have a feature that is different from the PD, where mixing of S and T shared by focal and opposing players can obtain a higher payoff than mutual Cs (or R), which is the best cooperative solution in PD (R-reciprocity). This unique feature of obtaining a high payoff by sharing S and T is called ST-reciprocity. In terms of direct reciprocity, ST-reciprocity seems to be as important as mutual cooperation in PD.

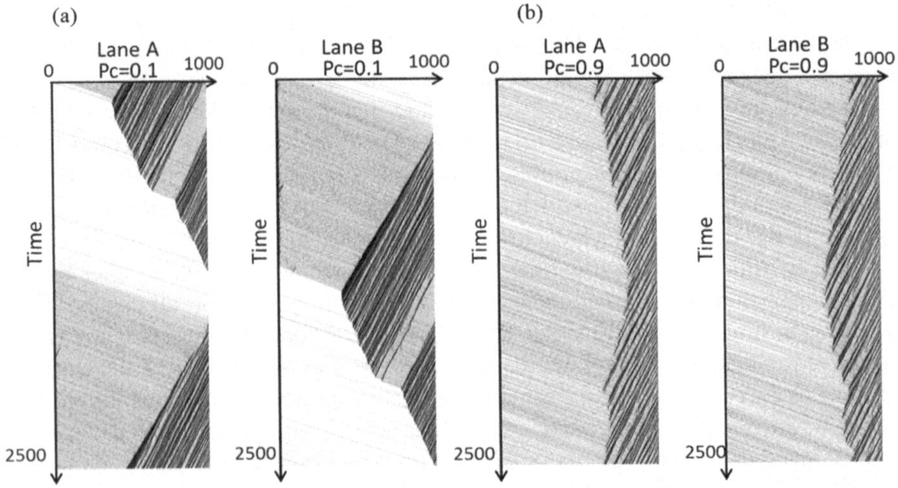

Fig. 3.25 Spatiotemporal diagrams on both routes for (**a**) $P_c = 0.1$ and (**b**) $P_c = 0.9$ when $r = 0.1$ and $\rho = 0.2$, which is indicated by a dotted line in Fig. 3.24a. The black dot indicates a cell occupied by a vehicle

misleading and inappropriate information given to defectors, who are the majority in this setting. Thus, at least some defectors should select another route despite given information. If the information is based on the travel time of someone who has most recently passed each of the two routes, the information does not represent the situation for each alternative route, because the information is respectively observed in certain past moments. Thus, this situation causes error and time-delay when compared with appropriate information that can lead to SM, where a 50–50 share should be between routes A and B. However, some ITSs may rely on this kind of information in practice. In the cooperative case ($P_c = 0.9$), panel (b), a time fluctuation, as observed in the $P_c = 0.1$ case, does not occur, despite the small emerging jam just before the exit. Thus, we can conclude that a huge traffic jam never occurs in routes A and B in the cooperative case ($P_c = 0.9$).

Figure 3.26 shows the dilemma-occurring condition and its strength when varying the quick-start parameter, r, and traffic density, ρ. Referring to our previous study,[31] the definition of the dilemma strength in this study is

$$\xi = \frac{q_{\max} - q_{NE}}{q_{\max}} \cdot \frac{q_{\max} - q_{\min}}{q_{\max}}, \tag{3.31}$$

where q_{max}, q_{NE}, and q_{min} indicate the maximum flux (social-maximum flux), flux at NE, and social-minimum flux, respectively, when varying P_c. Note that $0 \le q_{\min} \le q_{NE} \le q_{\max} \le 1$, which guarantees that ξ is always positive. The first part of the current definition, $\frac{q_{\max} - q_{NE}}{q_{\max}}$, is the same as the previous definition presented in Nakata et al.,[31] which indicates that social inefficiency should be evaluated by the

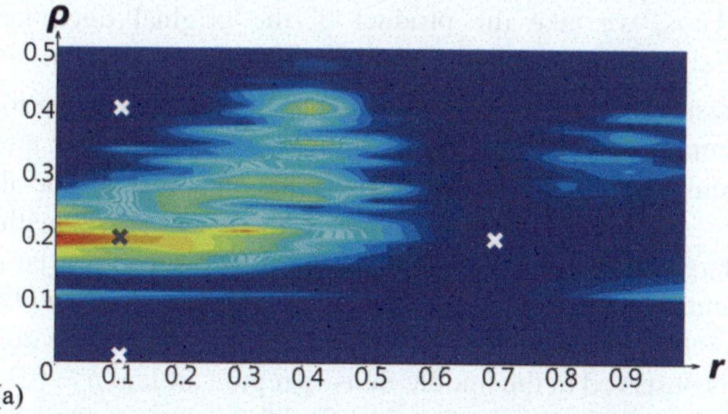

(a)

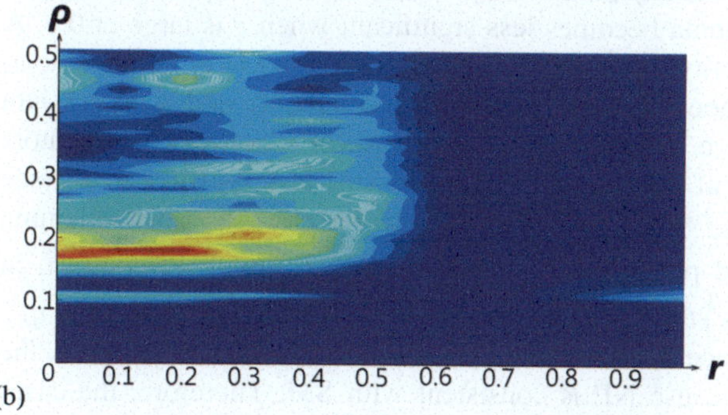

(b)

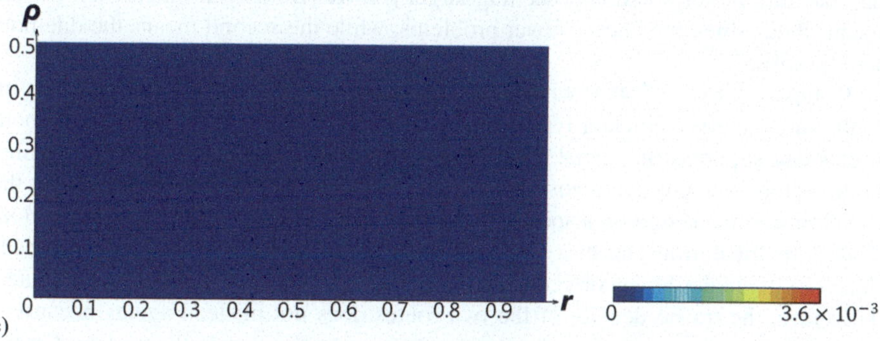

(c)

Fig. 3.26 Normalized dilemma strength, ξ, on a two-dimensional plane of the quick-start parameter r and traffic density ρ. The information given to defectors is the (**a**) travel time, (**b**) average velocity, and (**c**) density

ratio of the gap between social max (SM) and Nash equilibrium (NE) to SM. The first part can be large when q_{max} is small. However, when the gap between social maximum and minimum is sufficiently small, the dilemma is not a challenge, even if $\frac{q_{\max}-q_{NE}}{q_{\max}}$ indicates a large dilemma strength. To compensate for this, we have added the second part, $\frac{q_{\max}-q_{\min}}{q_{\max}}$, which is the normalized flux gap between social maximum and minimum. Thus, we take the product of the original dilemma strength, $\frac{q_{\max}-q_{NE}}{q_{\max}}$, and its social meaning (social impact), $\frac{q_{\max}-q_{\min}}{q_{\max}}$.

Figure 3.26a shows the result of taking travel time as the information given to defectors. The Chicken dilemma only appears when middle ρ and smaller r are assumed. A small traffic density allowing the free-flow phase causes no social friction between cooperators and defectors because all of them can travel with V_{max} (Fig. 3.24d). Also, a large traffic density that puts the flow field into a jam phase contains a large stop-and go wave, causing less social friction among C and D agents, because they are involved with large jams that result in vehicles driving slowly. The social dilemma is observed in the middle-density region around $\rho = 0.2$, which is consistent with the density transferring from the free-flow phase to the jam phase. Here, the social dilemma becomes less significant when r is large at 0.6. A larger r means that the majority of vehicles, regardless of either C or D, are able to quick-start, resulting in smooth acceleration and deceleration and a lower jam frequency. As a consequence, a lower difference in payoffs between cooperators and defectors in the region of P_c makes the social payoff less sensitive to P_c (see Fig. 3.24c). Therefore, $\frac{q_{\max}-q_{\min}}{q_{\max}}$ is small. Thus, a larger r causes small dilemma strength because the second part in Eq. (3.1), $\frac{q_{\max}-q_{\min}}{q_{\max}}$, is small. In contrast, in Fig. 3.24b, a larger ρ causes small dilemma strength. In nutshell, small $q_{\max} - q_{NE}$ and large $q_{\max} - q_{\min}$ reduce the dilemma strength. Because$q_{\max} - q_{NE} \cong 0$, the dilemma will be solved, because NE is consistent with SM. Therefore, the cases approaching larger r and approaching larger ρ differ from each other: the first one means that a dilemma causes fewer problems, while the second means the dilemma will be solved.

Comparing Fig. 3.26b, c with Fig. 3.26a, the average velocity of all vehicles on both routes causes a chicken-type social dilemma, whereas the traffic density of both routes can suppress the social dilemma over the entire region. Unlike travel time, both average velocity and density are global information. As mentioned above, the travel time is measured on a specific vehicle, which is local, not global information. Therefore, travel time causes a social dilemma because "local" information misleads defectors to make an incorrect choice that prevents sharing between two routes. Therefore, the traffic density of the two routes does not mislead agents because it represents the most important factor that causes a traffic jam in the system. On the other hand, the average velocity, despite being global, not local information, does not work well. For example, route A being metastable while route B is freely flowing means that the global traffic density of route A is much higher than that of route B. Thus, the number of vehicles driving on route A is much higher than that of route B. However, the average velocities on both routes are approximately the same. In this situation, "average velocity" does not provide useful information for selecting one of the two routes, namely route B. When providing traffic density as information, agents who depend on the information can select fairly as observed in Fig. 3.26b, c. Therefore, the average velocity does not solve the Chicken-type dilemma as the density can.

3.2.3 Summary

To determine whether route selection has a social dilemma behind its flow field, we established a CA model based on the revised S-NFS model. We attempted to determine what information can cause or avoid a dilemma during route selection.

When the travel time of alternative routes is provided to agents who rely on information (defectors), we found that a multiplayer Chicken-type dilemma occurs in the moderate-traffic-density region when fewer vehicles are considering a quick-start effect. We provide the dilemma strength of the multiplayer Chicken game when varying the traffic densities and quick-start-adoption rates.

However, this chicken-type social dilemma can be eliminated when density information is provided. Interestingly, when the average velocity, which, like the density, is qualitatively global, is given, the social dilemma cannot be solved.

Density information can be given to agents to relieve social dilemmas and improve social efficiency.

3.3 Social Dilemma in 2-Body Problem of Action and Reaction During Lane-Changing

There is no doubt that a driver's decision-making process, appearing in some cases through stochastic deviations, adds significant complexity to traffic-flow systems.[39] This insight motivates us to dovetail a traffic-flow model with evolutionary game theory that can emulate human decision-making in a mathematical framework, with the aim of understanding traffic flow composed of human–vehicle-interactive systems as real complex entities. This is exactly what this book poses as its main theme. In a sequence of works[29,30,31,32], we found that drivers' desire to go first triggers frequent lane-changing actions, causing turbulence in the original traffic flow, letting traffic efficiency decrease from a social optimum where no one tries a lane-change, thus, bringing the social-dilemma structure to an end. This seems scientifically interesting because traffic flow, which is thought to be an application supported by fluid dynamics and regarded as a hard-core "science & engineering" issue, describes a social dilemma that arises from human intention. However, one plausible question must be posed here. It is that our previous model dealing driver's decision whether taking lane-changing or not presumes a simple unilateral relation of a lane-changing vehicle affecting on his following vehicle, does not consider a chain of action – reaction introduced by the two neighboring vehicles. In reality, a focal vehicle may or may not hamper traffic by accelerating (decelerating) if a preceding vehicle is

[39]There have been many related works, one of which has been discussed in the previous section. Here, let us suggest another: Zhang and Chen (2012).

trying to change from the neighboring lane to the present one, so as to drive its neighbor's payoff down (up). Both spiteful and assisting attitudes are commonly observed in human behavior, as has been studied amongst evolutionary game theorists under the name "sanction"[40] Obviously this two-body relation affects traffic-flow dynamics by adding more complexity to the system. Elucidating this particular phenomenon would be another challenge contributing to complexity science as well as non-linear dynamics.

Based on this background, we build a new traffic-flow model incorporating evolutionary game theory to consider the two-body problem between a lane-changing vehicle and a vehicle that hampers it, and to explore whether or not any specific social-dilemma structures can be observed under different flow phases stipulated by traffic density.

3.3.1 Model Setup

We presume a general traffic flow on a highway with a one-directional two-lane system, where a focal vehicle may change lanes as long as it is beneficial and safe for itself. We also account for the situation where a vehicle may hamper (assist) a lane-change tried by another vehicle by means of accelerating (braking). To ensure our objective, we adopt a revised S-NFS model with a lane-changing sub-model considering the velocity difference of two successive vehicles, both of which were discussed in Sect. 3.1. Obeying our previous studies[29,32], we presume cyclic boundary conditions of which the system length is L (=500), because it is able to preserve traffic density in the domain. We also introduce a framework supported by evolutionary game theory to reproduce the two-body problem. In the following text, we call a human-driven vehicle an agent. The number of agents generated in the system controls traffic density.

3.3.2 Agent's Strategy

We impose a 2-dimentional strategy system on all agents, regardless of whether they change lanes if possible (defection; D) or never change lanes (cooperation; C) and whether they hamper other agents' lane-change (accelerating; A) or assist it (braking; B). Hence, an agent's strategy is described by two-capital letters, {CB, CA, DB, DA}, which is featured with the discrete four strategies as a whole.

---

[40]With respect to "sanction" in an evolutionary game, there has been a rich stock of previous works. A representative work is: Schoenmakers et al. (2014).

3.3.3 Vehicle Dynamics

We adopt a revised S-NSF model.

For the lane-changing sub-model, based on what was discussed in Sect. 3.1, we consider both an incentive criterion and a safe criterion.[22] See Fig. 3.27 for the nomenclature.

$$\text{Incentive criterion}: \; gap_p^f \le v_i^{(p)} - v_{i+1}^{(p)} \wedge \; gap_n^f > v_i^{(p)} - v_{i+1}^{(p)} \wedge \; gap_n^f > 0, \tag{3.32a}$$

$$\text{Safe criterion}: gap_n^b > v_{i-1}^{(n)} - v_i^{(p)}. \tag{3.32b}$$

According to Kukida et al.,[22] although the safe criterion allows more frequent lane-changing, Eq. (3.32) as a whole provides a likely sub-model for dealing with lane-changes. The conventionally accepted Safe criterion, $gap_n^b > v_{i-1}^{(n)} - 1$ or $gap_n^b > V_{\max}$, is stricter to avoid rear-ending. By contrast, Eq. (3.32b) is more relaxed. Although a real context might be in-between those two ideas, we presume Eq. (3.32b) in this study.

Let us consider an event as below. An agent (A in Fig. 3.27) with strategy D* (hereafter, * indicates a wild-card) meets with the incentive criterion (Eq. (3.32a)) and turns on a blinker to notify his following agent (B) of his intention to change lanes.

Subsequently, agent B in the neighboring lane of the focal agent (A) reacts according to his strategy, either *B or *A, only when the distance to the focal agent is below D_{cri}. We define D_{cri} as (V_{max} + 1) +1 as the critical distance below which agent B's reaction is triggered by A. V_{max} means the maximum velocity in the

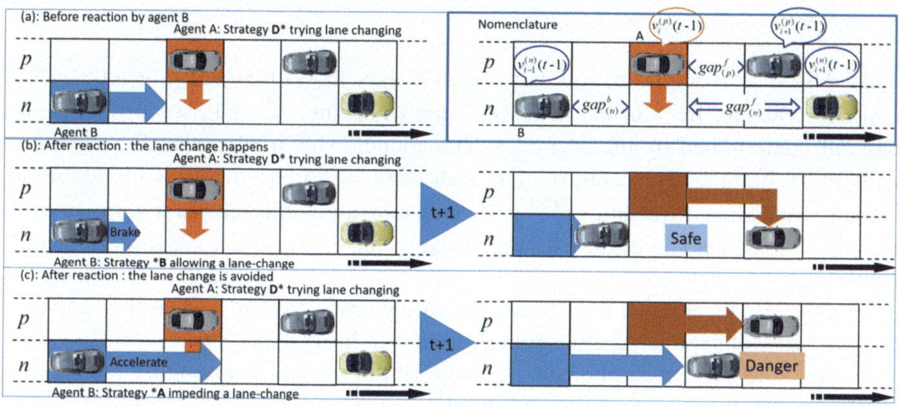

Fig. 3.27 Scheme of the proposed model. (**a**) configuration at time-step t; (**b**) transition from t to t + 1 when lane-changing is assisted by Agent B; (**c**) lane-changing is impeded. The highlighted box gives the nomenclature for the mathematical formulation

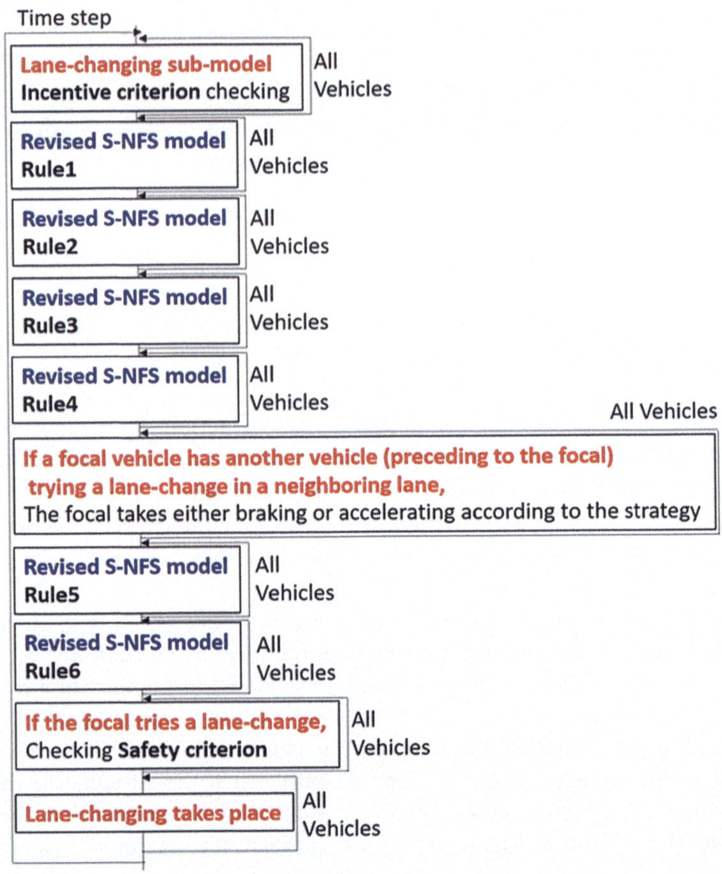

Fig. 3.28 Flowchart of the model combining the revised S-NFS model with our changing sub-model

revised S-NFS model. If this follower (agent B) has *B as his strategy, he reduces his temporal velocity by 1 in the next time step (meaning $v_i^{(4)} - 1$; see Eq. 3.25), as basically determined by the revised S-NFS model. This implies that B cooperates with A by braking. By contrast, if B's strategy is *A, he modifies his temporal velocity by adding 1 (meaning $v_i^{(4)} + 1$). This may allow this temporal velocity over V_{max} by +1, which should be allowed because this spite action takes place at this instance.

Finally, the focal agent (A) checks the safety criterion. Only when it is satisfied does A really change lanes after going forward in his original lane.

Figure 3.28 presents an holistic structure of the model incorporating the revised S-NFS model, the lane-changing sub-model and the sub-model of reaction.

3.3.4 Simulation Setup

In a series of simulations, we vary traffic density by controlling the number of agents in the domain. We also vary strategy fractions among agents ($f_{CB}, f_{CA}, f_{DB}, f_{DA}$) with an increment of $\Delta f = 0.1$. In a single simulation episode, we randomly place agents at the beginning. We obtain the traffic flow, average velocity of each strategy, number of lane-changing events and so forth after the traffic-flow field shows a quasi-equilibrium, being less influenced by the initial configuration. We perform 100 independent realizations starting from different initial configurations to obtain robust statistical results.

3.3.5 Framework of the Evolutionary Game

Let us measure each strategy's payoff with the average velocity of agents having strategy *I*, *V*(*i*) ($i \in$ {CB, CA, DB, DA}), at a quasi-equilibrium, presuming a combination of strategy fractions of ($f_{CB}, f_{CA}, f_{DB}, f_{DA}$). Let $f_{CB} + f_{CA} + f_{DB} + f_{DA} = 1$. Using replicator dynamics, we can define a time-derivation of the strategy of the fraction of drivers adopting a given strategy at a certain time-step as bellow;

$$\Delta f_i = \left[V(i) - \mathrm{Ave}[V(j)]_{j \in \{\mathrm{CB,CA,DB,DA}\}} \right] \cdot f_i. \tag{3.33}$$

Operator "Ave[]" takes an average of *V*(*i*) over the four strategies. The time-derivation as a whole; described by $\mathbf{\Delta f} \equiv (\Delta f_{CB} \;\; \Delta f_{CA} \;\; \Delta f_{DB} \;\; \Delta f_{DA})$, indicates how agents modify their strategy if it can be altered, which is provided by each vector in trajectory diagram of the strategy, Fig. 3.30, discussed later. Here, let $\sum \Delta f_i = 0$.

3.3.6 Results and Discussion

Figure 3.29 shows a so-called fundamental diagram indicating a relation between traffic density and traffic flux, regarded as the most important result in the field of traffic-flow analysis. Since panel (a) presumes $f_{CA} = 1$, the fundamental diagram must recover a default situation where all vehicles keep their respective lanes with no lane-changing at all (thus, results for $f_{CA} = 1$ must be consistent with f_{CB}.) Hence, we can confirm a free-flow phase, appearing at relatively low traffic densities with clear relation between density and flux of which slope means V_{max}, and jam phase, appeared in relatively high-density region. We also confirm there is a plot-scattering region between the free-flow and jam phases containing a meta-stable phase.

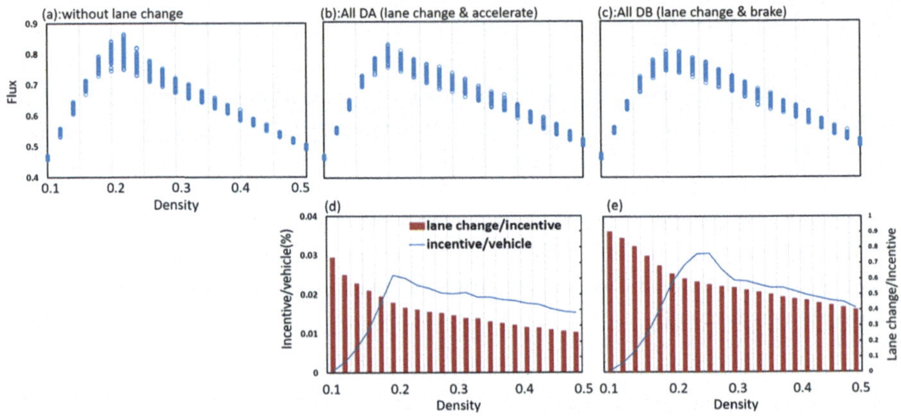

Fig. 3.29 Top (panels (**a** – **c**): fundamental diagram (relation between traffic density and flux). We presumed: (**a**) $f_{CA} = 1$, (**b**) $f_{DA} = 1$, and (**c**) $f_{DB} = 1$. Bottom (panels (**d**) and (**e**)): frequency of lane-changing action taken by agents (blue line) and its success rate (red bar). Panel (**d**) is for $f_{DA} = 1$ and (c) is for $f_{DB} = 1$

In panel (b) we presume that $f_{DA} = 1$, the most "egocentric" setting, whereby every agent tries lane-changing but never allows others to do so. Obviously, an inefficient traffic flow as compared with panel (a) is inevitably observed. In fact, the peak flux appears lower than that of panel (a). Panel (b) also shows the frequency of lane-changing actions taken by agents and its success rate. This explains that the meager peak flux vis-a-vis panel (a) results from the flow turbulence encouraged by lane-changing. In panel (c) we presume $f_{DB} = 1$. This setting is preferable, from a social-welfare viewpoint, to panel (b), because all agents cooperate for each other's lane-changing. Ironically, the peak flux falls below that observed in panel (b). Frequent introduction of braking to assist other's lane-changing deteriorates social efficiency rather than ameliorating it.

Figure 3.30 illustrates a compositive result that captures what evolutionary path can be observed. The lower central panel (panel (d)) shows relations between max-flux, q_{max} (i.e., the social-optimum flux), min-flux, q_{min}, flux at Nash equilibrium, q_{NE}, dilemma strength, η, and κ vs traffic density, ρ, when the strategy fraction $(f_{CB}, f_{CA}, f_{DB}, f_{DA})$ is varied. According to previous works[24,29,31,32], we define the dilemma strength, η, as below;

$$\eta = \frac{q_{\max} - q_{NE}}{q_{\max}}. \tag{3.34}$$

q_{max}, q_{min}, and q_{NE} mean the following. Fixing traffic density at a certain ρ but varying strategy fraction; $(f_{CB}, f_{CA}, f_{DB}, f_{DA})$, we can find maximum and minimum fluxes, namely q_{max} and q_{min}, respectively. The strategy fraction at Nash equilibrium can be found by chasing flow lines suggested by the vectors shown in the upper three panels. Let us note that the strategy fraction at Nash equilibrium is pointed out by

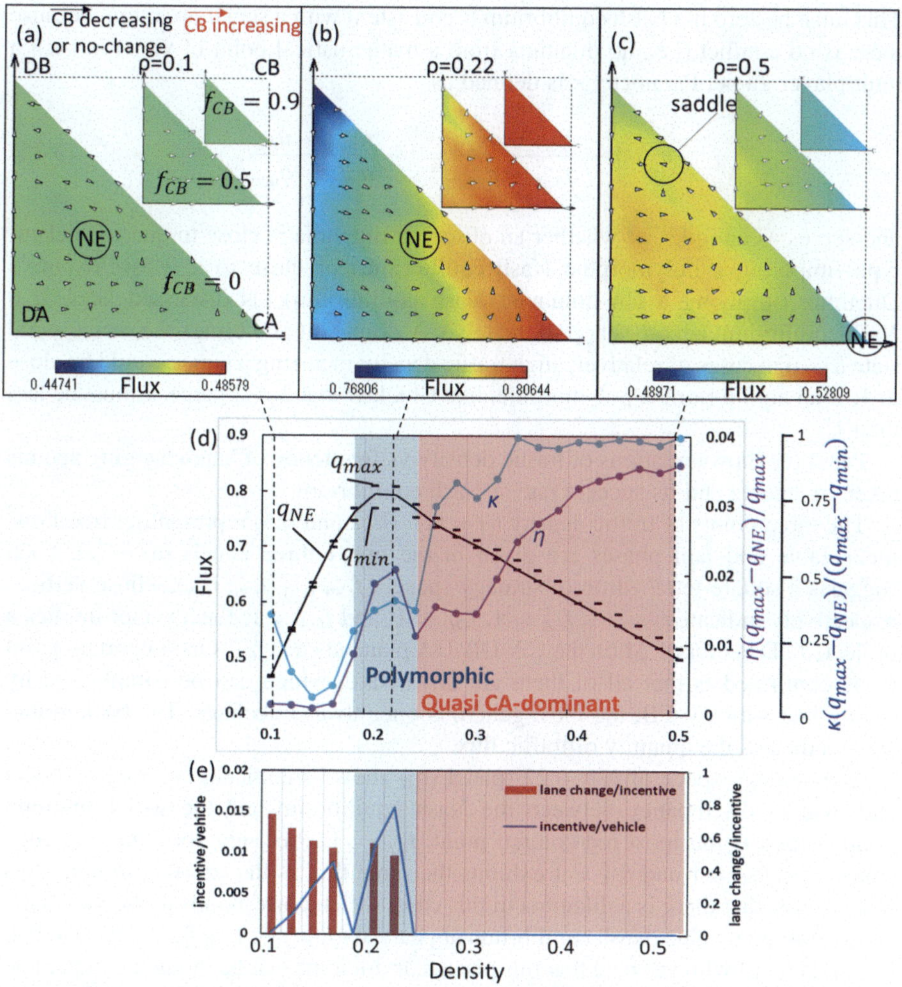

Fig. 3.30 (**a–c**); strategy space of ($f_{CB}, f_{CA}, f_{DB}, f_{DA}$) at traffic densities of (**a**) 0.1, (**b**) 0.22, and (**c**) 0.5. An arrow indicates a projected 2D vector of $\mathbf{\Delta f}$ on the CA-DB-DA plane at $f_{CB} = 0$, 0.5, and 0.9. Black and red respectively indicate f_{CB} decreasing and increasing. A large open circle indicates Nash equilibrium. Panel (**d**) gives a fundamental diagram at Nash equilibrium with maximum and minimum fluxes, κ, and η. Panel (**e**) shows frequency of lane-changing actions taken by agents (blue line) and its success rate (red bar). Thin blue- and red-highlighted regions indicate Chicken-type (polymorphic) and PD-type (quasi CA-dominant) dilemmas, respectively

each open circle in each upper panel ((a) – (c)). The Nash equilibrium implies a final social state in this dynamical system allowing the strategy fraction to be changed obeying Eq. (3.33) at a certain traffic density. Thus, q_{NE} indicates the traffic flux at Nash equilibrium. The dilemma strength, defined in Eq. (3.34), indicates how Nash equilibrium differs from social maximum, potentially implying a social inefficiency.

This must be zero if a Nash equilibrium is consistent with a social optimum, because there is no conflict (i.e., no dilemma from a mathematical point of view) in such a multiplayer game. Parameter κ is defined as

$$\kappa = 1 - \frac{q_{SO} - q_{NE}}{q_{\max} - q_{\min}} = 1 - \frac{q_{\max} - q_{NE}}{q_{\max} - q_{\min}}, \tag{3.35}$$

and serves as an index of whether an observed dilemma is close to quasi-Chicken-type (implying a polymorphic Nash equilibrium) or close to a quasi-Prisoner's Dilemma (implying a CA-dominant Nash equilibrium), as discussed later. If a Nash equilibrium becomes approximately CA-dominant and brings the worst social state (q_{min}) in cases of relatively high traffic density incurring a jam, κ would be close to 1. If an equilibrium is polymorphic, like Chicken in a 2 $\times$ 2 game, κ must be less than 1.

Panel (e) shows relations of traffic density vs frequency of lane-changing actions taken by agents and its success rate at Nash equilibrium.

The three points of traffic density ($\rho = 0.1$, 0.22 and 0.5) representing free-flow, meta-stable and jam phases are given in the upper three panels (a) – (c). Each suggests a square-tetrahedron of strategy space, ($f_{CB}, f_{CA}, f_{DB}, f_{DA}$), whose vertices respectively indicate $f_{CB} = 1, f_{CA} = 1, f_{DB} = 1$, and $f_{DA} = 1$. Each vector implies a projected 2D vector of $\mathbf{\Delta f}$ on the CA-DB-DA plane at each f_{CB}. One important point to be confirmed is that all of them are downward, which can be paraphrased by saying that strategy CB, the most generous one, always declines. The background color indicates the quantity of traffic flux.

The strategy space of $\rho = 0.22$ (panel (b)) shows significant coloring diversity and evident discrepancy between the Nash equilibrium and the social-optimum point, which is the most red-colored point, $f_{CB} = 1$. Precisely speaking, the edge connecting $f_{CB} = 1$ and $f_{CA} = 1$ exhibits the same flux as the social optimum. This fact ensures that there is a dilemma in this case, which corresponds to the fact that η shows non-zero value. Nash equilibrium appears at ($f_{CB}, f_{CA}, f_{DB}, f_{DA}$) = (0, 0.4, 0.4, 0.2). It is worthwhile to note that this particular dilemma can be thought comparable to a Chicken-type dilemma (e.g., Tanimoto and Sagara[34] and Wang et al.[35]) in the case of a 2 $\times$ 2 game. This is because Nash equilibrium is a polymorphic inner equilibrium where all strategies except for CB co-exist, and q_{NE} is not close to q_{min}, unlike the case of $\rho = 0.5$ discussed in the following paragraph.

Qualitatively paraphrasing, a traffic-flow field potentially belonging to a meta-stable phase may suffer from a large social dilemma. From the standpoint of traffic-flow physics, a meta-stable phase is very vulnerable. Because of specific vehicle configurations such as "platoon running", any subtle perturbation may trigger disruption of a meta-stable phase, significantly reducing traffic flux. What we have shown here suggests not only conventional causes for the perturbation, such as deviation of the driver's manner of operation or stochastic deviation of the surrounding environment (i.e., external or mechanical factors), but also social-dilemma mechanisms that affect driver psychology (i.e., an internal factor).

It is worthwhile to confirm that in this traffic-density region (highlighted blue in panel (d)), strategy C* really means a cooperative strategy, since its action of self-prohibiting lane changes ensures that a social-optimum state appears on the edge connecting $f_{CA} = 1$ and $f_{CB} = 1$. But the real end point of a social evolution, Nash equilibrium, leads to a certain polymorphic state with co-existence of strategies other than CB. This seems analogous to the pure-Chicken-type dilemma in a 2×2 game, where *R*-reciprocity (Tanimoto and Sagara[36] and Wakiyama and Tanimoto[37]) ensures a mutually equal pareto optimum (that is social optimum). In a 2×2 game, mutual cooperation (C-C; called Reward; *R*), mutual defection (D-D; called Punishment; *P*), a focal player cooperating and an opponent defecting (C-D; Sucker; *S*), and a focal player defecting and an opponent cooperating (D-C; Temptation; *T*) are possible. In a pure-Chicken-type dilemma, mutual cooperation realizes a social optimum (Thus, it is called *R*-reciprocity), but Nash equilibrium appears at a polymorphic state (inner equilibrium). To this end of the discussion so far, let us call this type a polymorphic game bringing Chicken-type dilemma.

Next, let us consider the case with $\rho = 0.5$ (panel (c)). The dilemma strength in this case is clearly non-negligible and is larger than that at $\rho = 0.22$. Nash equilibrium appears at $(f_{CB}, f_{CA}, f_{DB}, f_{DA}) = (0, 1, 0, 0)$, which is a vertex of $f_{CA} = 1$, the all-CA state. Unlike $\rho = 0.22$, this particular dilemma seems analogous to the PD in 2×2 games, because its Nash equilibrium is the pure strategy point, its vertex is $f_{CA} = 1$, and q_{NE} is close to q_{min}, as observed at $f_{CA} = 1$. This fact is qualitatively supported by κ shown in panel (d), which is close to 1 in the higher-traffic-density region, highlighted red in panel (d). In a 2×2 PD game, *R*-reciprocity can provoke a mutually equal pareto optimum (like a pure-Chicken game), and mutual defection induces a Nash equilibrium. Let us call this type a quasi-CA-dominant game bringing a PD-type dilemma.

In this case, strategy CA (highlighted red in panel (d)) is destructive, or most "egocentric", despite the presence of "C". Observing the frequency of lane-changes taken by agents and its success rate at Nash equilibrium, we notice that there is no chance of making a successful lane change (see panel (e). This fact implies that even if some defective agents try changing lanes, most trials will be prevented by the majority of agents, CA; thus, those who are defective cannot have an advantage over CA agents, and would ultimately be eliminated from the population. Whenever a CA agent successfully hampers another's lane-change, this event pushes his payoff up due to temporal acceleration. This seems ironic in a sense, because an action introduced by a CA agent results in working as a significantly egocentric manner as a consequence despite his label of cooperative, C*. Also, a CA agent does not contribute to average out if there is a local density gulf among two neighboring lanes, because he always keeps to his original lane. In terms of holistic social efficiency when a global traffic density is relatively large, a moderate lane-changing frequency may rather improve global traffic flux by dissolving a local vehicle cluster appearing in one single lane. For these reasons, CA works as an absolutely defective strategy in terms of global social efficiency.

A crucial point to be addressed here is that the dilemma structure observed at a middle density around a meta-stable phase (Chicken-type dilemma) is thought to shift to an alternative dilemma structure (PD-type dilemma), as observed in a higher-density jam phase, as suggested by the lower panel (d). Also, shifting from the Chicken-type dilemma to the PD-type dilemma, the substantial meaning of the first letter to express strategy, “C”, entirely flips, shifting from “cooperative” to “defective”.

Consider $\rho = 0.1$ (panel (a)). We find that this case in the free-flow phase has less color diversity and is almost mono-tone, although chasing flow lines leads to a certain Nash equilibrium. Thus, the flux discrepancy between Nash equilibrium and a social optimum is negligible. Hence, this case shows only slight dilemma strength, much less than the previous two cases, and can be neglected. The Nash equilibrium causes strategies other than CB to co-exist, implying that some defective agents who try to change lanes can exist. This does not contradict social welfare, because an appropriate lane-changing pushes traffic efficiency up without harming others due to low traffic density.

3.3.7 Summary

We built a new CA model dovetailed with the concept of evolutionary game theory to quantify how a driver’s spiteful behavior to hamper others’ lane-changing causes a social dilemma in a realistic traffic flow. We aim to take account of a two-body relation between a vehicle changing lanes and a vehicle intending to hamper/assist it as a reaction to the lane-change.

Our result reveals that the complex social system that allows four strategies (intending to change lanes or not, and intending to hamper or help others’ lane-changes) contains a social-dilemma structure. Interestingly, this structure can feature either a polymorphic equilibrium like Chicken or almost single-strategy dominance like the PD depending on the traffic density.

The first dilemma appears in a middle traffic density and takes place when the flow field potentially belongs to the meta-stable phase. Inversely speaking, this social-dilemma mechanism, brought by lane-changing as well as hampering/assisting a lane-change, results in breaking-platoon-running configurations in a flow field, leading to corruption of the meta-stable phase associated with a reduction of traffic flux. This type of flow inherently contains a Chicken-type dilemma.

The second dilemma takes place in accordance with the jam phase. This seems ironic, because a high traffic density makes a CA agent, the social majority, act as the most egocentric (most defective) for the following two reasons; (1) hampering others’ lane-changes, thereby preventing them from dissolving a locally skewed density among two neighboring lanes and bringing a benefit resulting from a temporal acceleration to the focal agent, and (2) never trying to change lanes at all, and thus not contributing to dissolving local vehicle clusters placed in one of the two lanes.

This fact confirms that this traffic-flow system allows the dilemma structure to change from Chicken-type to PD-type phases as traffic density increases, which is qualitatively consistent with instinct phase change x of the flow field from a meta-stable phase to a jam phase.

Unlike our previous works[29,30,31,32], which only observed PD-type social dilemmas behind traffic flows when considering lane-change intention, the current work finds there to be a rich complexity in terms of social dilemmas behind realistic traffic flows. This is attribute to the fact that we considered the two-body problem, namely an action–reaction chain in agents' relations.

References

Arthur, W. B.; Inductive reasoning and bounded rationality, *American Economic Review* **84**, 406–411, 1994.

Bando, M., Hasebe, K., Nakayama, A., Shibata, A., Sugiyama, Y.; Dynamical model traffic congestion and numerical simulation, *Physical Review E* **51 (2)**, 1035, 1995.

Fukuda, E., Tanimoto, J., Iwamura, Y., Nakamura, K., Akimoto, M.; Field measurement analysis to validate lane-changing behavior in a cellular automaton model, *Physical Review E* **94**, 052209, 2016.

Fukui, M., Ishibashi, Y.; Traffic flow in 1D cellular automaton model including cars moving with high speed, *Journal of the Physical Society of Japan* **65 (6)**, 1868–1870, 1996.

Gundaliya, P. J., Mathew, T. V., Dhingra, S. L.; Heterogeneous traffic flow modelling for an arterial using grid based approach, *Journal of Advanced Transportation* **42 (4)**, 467–491, 2008.

Haberman, R.; Mathematical models: Mechanical vibrations, population dynamics, and traffic flow, An introduction to applied mathematics, Society for Industrial Applied Mathematics (Philadelphia), 1977.

Jiang, R., Wu, Q., Zhu, Z.; Payne, H. J.; Full velocity difference model for a car-following theory, *Transportation Research Pert B: Methodological* **36 (5)**, 405–419, 2002.

Kerner, B. S.; Introduction to Modern Traffic Flow Theory and Control, Springer, 2009.

Kerner, B. S., Klenov, S. L.; Phase transitions in traffic flow on multilane roads, *Physical Review E* **80**, 056101, 2009.

Kokubo, S., Tanimoto, J., Hagishima, A.; A new Cellular Automata Model including a decelerating damping effect to reproduce Kerner's three-phase theory, *Phyisica A* **390 (4)**, 561–568, 2011.

Kukida, S., Tanimoto, J., Hagishima, A.; Analysis of the influence of lane changing on traffic-flow dynamics based on the cellular automaton model, *International Journal of Modern Physics C* **22 (3)**, 271–281, 2011.

Lan, L. W., Chiou, Y.-C., Lin, Z.-S., Hsu, C.-C.; A refined cellular automaton model to rectify impractical vehicular movement behavior. *Physica A* **388**, 3917–3930, 2009.

Nagel, K., Chreckenberg, M.; A cellular automaton model for freeway traffic, *Journal of Physics I France* **2**, 2221–2229, 1992.

Nakata, M., Yamauchi, A., Tanimoto, J., Hagishima, A.; Dilemma game structure hidden in traffic flow at a bottleneck due to a 2 into 1 lane junction, *Physica A* **389**, 5353–5361, 2010.

Nishinari, K.; A Lagrange representation of cellular automaton traffic-flow models, *Journal of Physics A; Mathematical and general* **34**, 10727–10736, 2001.

Nishinari, K., Takahashim D., Analytical properties of ultradiscrete Burgers equation and rule-184 cellular automaton, *Journal of Physics A; Mathematical and general* **31**, 5439–5450, 1998.

Nishinari, K., Takahashim D., Multi-value cellular automaton models and metastable states in a congested phase, *Journal of Physics A; Mathematical and general* **33**, 7709–7720, 2000.

O'Loan, O. J., Evans, M. R., Cates, M. E.; Jamming transition in a homogeneous one-dimensional systems :The bus route model, *Physical Review E* **58**, 1404 (1998).

Payne, H. J.; Models of freeway traffic and control, *Mathematical models of public systems Simulation Councils Proc. Ser.* **1**, 51–61, 1971.

Peng, G. H., Cai, X. H., Liu, C. Q., Cao, B. F.; A new lattice model of traffic flow with the consideration of the driver's forecast effects, *Physics Letters A* **375**, 2153–2157, 2011.

Sakai, S., Nishinari, K., Iida, S.; A new stochastic cellular automaton model on traffic flow and its jamming phase transition, *Journal of Physics A: Mathematical and General* **39**, 15327–15339, 2006.

Schoenmakers, S., Hilbe, C., Blasius, S. Traulsen, A.; Sanctions at honest signals The evolution of pool punishment by public sanctioning institutions, *Journal of Theoretical Biology* **356**, 36–46, 2014.

Tadaki, S., Kikuchi, M., Fukui, M., Nakayama, A., Nishinari, K., Shibata, A., Sugiyama, Y., Yoshida, T., Yukawa, S.; Phase transition in traffic jam experiment on a circuit, *New Journal of Physics* **15**, 103034, 2013.

Takayasu, M., Takayasu, H., 1/f noise in a traffic model, *Fractals* **1**, 860–866, 1993.

Tanaka, K., Nagatani, T., Masukawa, S.; Fundamental diagram in traffic flow of mixed vehicles on multi-lane highway. *Physica A* **387**, 5583–5596, 2008.

Tanimoto, J., Nakamura, K.; Social dilemma structure hidden behind traffic flow with route selection, *Physica A* **459**, 92–99, 2016.

Tanimoto, J., Sagara, H.; Relationship between dilemma occurrence and the existence of a weakly dominant strategy in a two-player symmetric game, *BioSystems* **90 (1)**, 105–114, 2007a.

Tanimoto, J, Sagara, H.; A study on emergence of Coordinated Alternating Reciprocity in a 2x2 game with 2-memory length strategy, *BioSystems* **90 (3)**, 728–737, 2007b.

Tanimoto, J, Fujiki, T., Wang, Z., Hagishima, A., Ikegaya, N.; Dangerous drivers foster social dilemma structures hidden behind a traffic flow with lane changes, *Journal of Statistical Mechanics: Theory and Experiment*, P11027, 2014a.

Tanimoto, J., Kukida, S., Hagishima, A.; Social dilemma structures hidden behind a traffic flow with lane changes, *Journal of Statistical Mechanics: Theory and Experiment*, P07019, 2014b.

Tanimoto, J., Fujiki, T., Kukida, S., Ikegaya, N., Hagishima, A.; Acquisition of the field measurement data relating to lane change actions, *International Journal of Modern Physics C* **26**, 1550072, 2015.

Tokihiro, T., Takahashi, D., Matsukidaira, J., Satsuma, J.; From soliton equations to integrable cellular automata through a limiting procedure, *Physical Review Letters* **76**, 3247–3250, 1996.

Wakiyama, M., Tanimoto, J.; Reciprocity phase in various 2×2 games by agents equipped with 2-memory length strategy encouraged by grouping for interaction and adaptation, *BioSystems* **103 (1)**, 93–104, 2011.

Wang, Z., Kokubo, S., Jusup, M, Tanimoto, J.; Universal scaling for the dilemma strength in evolutionary games, *Physics of Life Reviews* **14**, 1–30, 2015.

Whale, J., Bazzan, A. L. C., Klugl, F., Schreckenberg, M.; Decision Dynamics in a Traffic Scenario, *Physica A* **287**, 669–681, 2000.

Wolfram, S.; Statistical Mechanics of Cellular Automata, *Reviews of Modern Physics* **55 (3)**: 601–644, 1983.

Xiang, Z.-T., Xiong, L.; A weighted mean velocity feedback strategy in intelligent two-route traffic systems, *Chinese Physics B* **22 (2)**, 028901, 2013.

Xue, Y.; Analysis of the stability and density waves for traffic flow. *Chinese Physics* **11**, 1128–1134, 2002.

Yamauchi, A., Tanimoto, J., Hagishima, A., Sagara, H.; Dilemma game structure observed in traffic flow at a 2-to-1 lane junction, *Physical Review E* **79**, 036104, 2009.

Zhang, W., Chen, W.; Dilemma game in a cellular automation model with a non-signalized intersection, *European Physical Journal B* **85 (78)**, 2012.

Chapter 4
Social Dilemma Analysis of the Spread of Infectious Disease

Understanding and controlling the spread of infectious disease is a pressing issue for our society. Contemporary globally connected civilization is more at risk from various modern infectious diseases than classical ones such as pests, cholera, and tuberculosis. Over the last few years, pandemic outbreaks of highly virulent influenza, possibly related to avian flu, severe acute respiratory syndrome (SARS), and middle-eastern respiratory syndrome coronavirus (MARSE) have been a threat. Beyond this, the intentional spread of infectious disease, e.g., "bioterrorism", has come to be recognized as being just as dangerous as nuclear weapons. An infectious disease spreads on human social networks. Each individual can protect himself through several measures. Pre-emptive vaccination is thought to be most effective, although it incurs a partial cost to each individual. This brings about a social dilemma, because an individual may be able to rely on so-called "**herd immunity**" to avoid his own infection without himself being vaccinated. Also, besides vaccination, there may be several practical ways to protect against contagion, such as wearing a mask, keeping away from crowds, and self-isolation by leaving the home less often, which may be less costly and less effective than vaccination. In any case, there is a human-decision-making process regarding what steps should be taken, while the dynamics of infectious-disease spread can themselves be evaluated as a diffusion problem that has been well-studied in physics for many years. Thus, based on the concept of human–environment–social interaction, a basic-physics model for this diffusion problem that considers evolutionary game theory (EGT) may lead us to obtain some meaningful solutions that can be proposed to our society. Following the previous chapter explaining how EGT can be applied to traffic-flow analysis, this chapter describes this practical problem.

Human social networks are a central application of evolutionary game theory because the complexity of the underlying network serves as a key factor determining game equilibrium. The spread of an epidemic throughout such a network is mathematically described by percolation theory (see Fig. 4.1, which provides a schematic image of a 2D percolation model applied to the spreading dynamics of an infectious disease), which is an archetype of the physics of a diffusion processes. Vaccination,

J. Tanimoto, *Evolutionary Games with Sociophysics*, Evolutionary Economics and Social Complexity Science 17, https://doi.org/10.1007/978-981-13-2769-8_4

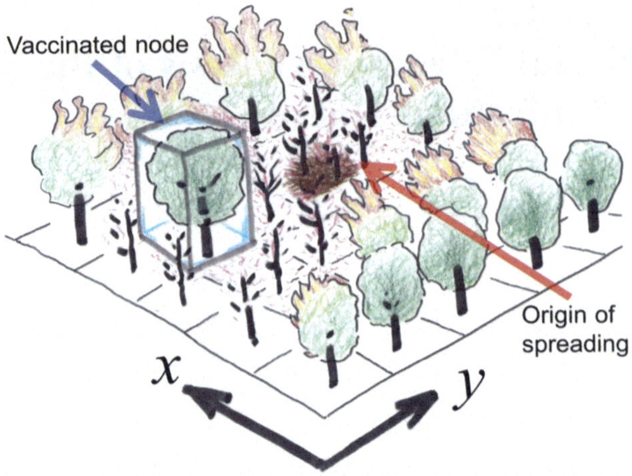

Fig. 4.1 The modeling concept of infectious-disease spread comes from the so-called "2D percolation model", commonly applied to 2D-diffusion problems such as wildfire. In this metaphor, firing is spreading a disease, and a perfectly vaccinated agent is represented by the tree in the box

which is driven by individual decision making, inhibits the spread of infectious diseases. In addition, if so-called herd immunity is established, a free-rider, who pays no cost for vaccination, can escape infection. Here, when we refer to vaccination "cost", we imply not only a direct cost, but also the potential risk of by-effects and psychological negative-costs brought about by vaccination. Obviously, there is a conflict between individual and social benefits; in short, a conflict between individual rational choices: trying to avoid vaccination, or everyone taking the vaccine to achieve the fair Pareto optimum, i.e., the best solution where everyone equally bears the cost to maintain public goods, namely the herd immunity in this context (see Fig. 4.2). This conflict is why we introduce evolutionary game theory into epidemiology; vaccination can be viewed as a game on a complex social network. This specific structure of the social dilemma has been called a **vaccination dilemma** and has been modeled in the framework of the **vaccination game**, explained later. However, vaccination is not an ultimate solution. In some cases, perhaps stochastically, the injection of a vaccine into a human body is not always able to establish immunity because of the imperfectness of the vaccination. Even in such an unwilling case, the vaccine may work to reduce the probability of being infected brought about by physical contact with people around the focal agent in his social network. An expected mathematical model should consider this kind of situation.

4.1 Epidemiological Model and Vaccination Game

Pre-emptive vaccination is one of the best public-health measures for preventing epidemics of infectious diseases and reducing morbidity and mortality.[1] However, most societies entrust vaccination to the autonomy of the individual: vaccination is

[1] Anderson and May (1991).

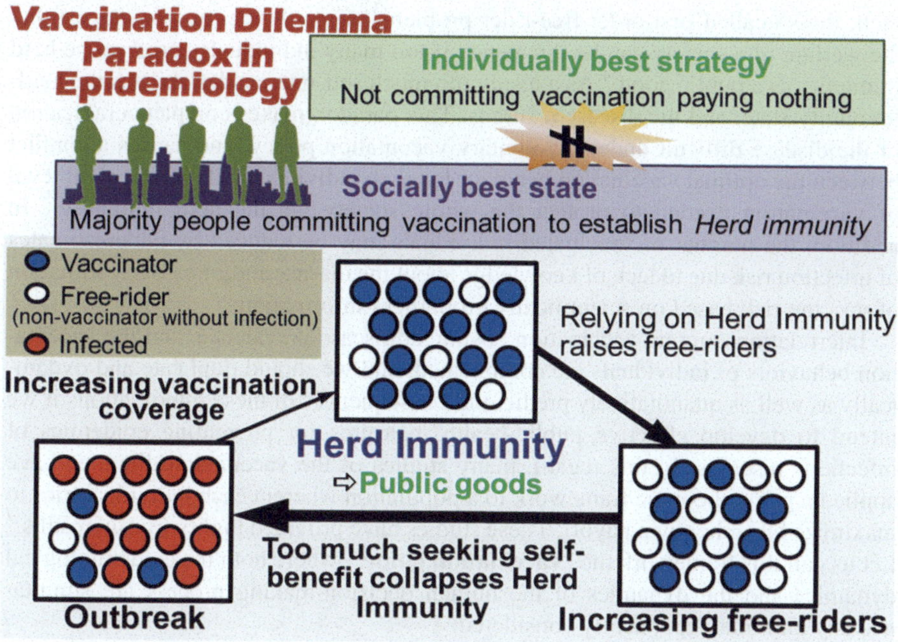

Fig. 4.2 Scheme of the social dilemma working behind the vaccination dilemma

usually voluntary, despite some national or local governments providing subsidies for it. Therefore, decision making at the individual level may be the result of a trade-off between protection and the perceived risks and costs of vaccination and infection. Furthermore, an individual's decision may be influenced by the vaccination behaviors of others.[2] The only example of a vaccination campaign that has completely eradicated a vaccine-preventable disease is smallpox, while cyclic (seasonal) epidemics of other infectious diseases, such as flu-like pathogens and influenza, remain a serious threat to humanity.

One major reason for the difficulty in eradicating vaccine-preventable diseases is related to an inherent vaccination dilemma, sometimes called the "**paradox in epidemiology**". As vaccination coverage increases over a population, the proportion of immunized individuals finally exceeds a critical level above which the disease can no longer persist; this point is called herd immunity, as mentioned above. Once herd immunity is attained, the remaining unvaccinated individuals are quite unlikely to become infected since they are indirectly protected by vaccinated individuals. Thus, unvaccinated individuals obtain benefits from the herd immunity without considering the perceived risks associated with vaccination, such as complications, side effects, and financial costs. There is less incentive for them to get vaccinated, and

[2]As representative works, we cite Chapman and Coups (1999, 2006), Basu et al. (2008).

then, the so-called first-order free-rider problem[3] arises. Some reports suggest that the welfare of a society can be threatened if too many individuals perceive the herd immunity as a public good.[4] As a result, too much self-interest destabilizes the herd-immunity state, and the disease resurges. This paradox makes complete eradication of the disease difficult under a voluntary vaccination policy, and causes a conflict between the optimal vaccination behavior for each individual and the sufficient level of vaccination needed to protect the whole society via the herd immunity.[5] In addition, the number of vaccinated individuals may be reduced by underestimates of infection risk due to lack of knowledge about the disease and/or by overestimation of vaccine risk based on scientifically groundless information.[6]

Interrelations among vaccination coverage, disease prevalence, and the vaccination behaviors of individuals are complicated, and we should duplicate and dynamically as well as quantitatively predict the consequences of these interrelations if we intend to develop effective public-health measures for preventing epidemics of infectious diseases. In this regard, many studies of the vaccination dilemma have applied a game-theoretic framework to a population wherein each individual tries to maximize his or her own payoff. These studies have provided highly fruitful results.[7] Let us call this framework the "**vaccination game**", where both the epidemiological dynamics and the dynamics of the human decision-making process are simultaneously and interdependently considered.

Some of the previous game-theoretic analyses of vaccination behavior have assumed a static game wherein individuals always act with perfect information about their probability of becoming infected. In reality, individuals cannot precisely know this probability. Moreover, the game should allow individuals to update their strategies through learning by imitating others who appear to have adopted more successful strategies. In this context, imitating others means adapting one's strategy based on one's own personal experience and information from media (the former and latter can be called active and passive information, respectively). Also, it is very likely that would be someone who acts opposite to what surrounding people do, because non-vaccination would cost a player nothing if his all neighbors were vaccinating.[8] As those, concerning how an individual updates his own decision,

[3]In 2 × 2 games, a defector who is harmful to cooperators is called a *first-order free-rider*. When a costly punishment scheme for defectors exists, there can be defined a strategy called the "masked good guy", who cooperates with others but never punishes defectors; such an individual is called a *second-order free-rider*. There is much literature on the second-order free-rider problem. For example, Olson (1965), Axelrod (1986), Yamagishi (1986).

[4]Asch et al. (1994).

[5]Although there are many good references on this issue, several representative ones are cited here. Cullen and West (1979), Fine and Clarkson (1986), Geoffard and Philipson (1997), Bauch et al. (2003), Bauch and Earn (2004).

[6]Jansen et al. (2003).

[7]There are many related references, but due to space limitations, we only cite the most representative here. Bauch (2005).

[8]Such behavior might be meaningful in the context of the minority game. See the following report.

not only imitating others; either his neighbors or media, but also drawing his specific decision based on the observation around him, there might be diverse ways in a real world. To describe this process explicitly, we should construct an appropriate model that combines mathematical epidemiological dynamics with game-theoretic dynamics, taking account of the various rules for strategy adapting. For example, Bauch constructed and analyzed a model that combines epidemiological dynamics with replicator dynamics of evolutionary game theory to capture the imitative behavior of individuals during outbreaks of diseases; he found that imitative behavior provokes periodic outbreaks of such diseases[7]. Vardavas et al. proposed an individual-level adaptive decision-making model that was inspired by a minority-game methodology.[9] By solving the model numerically and analytically, they showed that incentive-based vaccination programs are indispensable for controlling epidemics of infectious disease but that misuse of these programs may lead to a severe epidemic. These studies assumed that the population is homogeneously mixed and that individuals are fully rational in the sense that they make decisions to pursue maximum personal utility based on their perceived risks. Yet, in reality, there are always spatial structures for networks of both disease transmission and an individual's contacts, and any individual's behavior is not completely rational. Accordingly, Fu et al., for example, elevated a model to that of evolutionary game theory to explore the effects of individual adaptation behavior and population structure upon vaccination when a population is faced with an epidemic of an infectious disease.[10]

Let us revisit the term paradox in epidemiology in the context of a game-theoretical application. Any rational individual has a strong incentive to exploit the public good by free-riding on herd immunity. However, this incentive, wherein the individual pays nothing but still obtains a benefit, only works as long as the majority of the community spontaneously receive the vaccination. By contrast, if the majority disregards vaccination, then doing nothing is no longer a better option because infection is likely. In this case, spontaneous vaccination becomes the rational option. This difference implies that the best choice for an individual is to always adopt the strategy of the social minority; either free-ride when the herd immunity is well established or take the vaccination when most people neglect to do so. This situation obviously contains the structure of a minority game, as Vardavas pointed out[8]. A minority game,[11] originally defined as the El Farol Bar problem,[12] is a typical social dilemma that can be observed in many real situations. The most heavily concentrated applications are in financial markets. In a minority game, any individual has an incentive to adopt the strategy of the minority under any circumstance. This duality might be interpreted as a Chicken-type dilemma wherein the fair Pareto optimum is realized when two strategies coexist, as discussed in Chap. 2.

---

[9] Vardavas et al. (2007).

[10] Fu et al. (2011).

[11] A reader can consult with, for example; Challet et al. (2005).

[12] Brian Arthur (1994).

Based upon the abovementioned review, this section gives the fundamental frameworks of both the epidemiological model and the vaccination game.

4.1.1 SIR/V (SVIR) Model for an Infinite & Well-Mixed Population[13]

We start our discussion with the simple situation, for which dynamics can be formulated by a set of ordinary differential equations (ODEs). Before the discussion, let us introduce one assumption that is substantially important in relation to vaccination. So far, we have implicitly presumed that a vaccination brings an idealized perfect immunity whenever performed. In reality, for infectious diseases such as flu, measles, malaria, and HIV, vaccination does not work perfectly, giving rise to the concept of the "effectiveness" of a vaccination. This presumes a situation in which some vaccinated agents acquire immunity with effectiveness probability[14] e; meanwhile, the remaining agents fail to acquire immunity with probability $1 - e$. This can be likened to a lottery that pays out either 100% or 0% of the prize fund according to the probability of winning. Meanwhile, there are protective measures other than vaccination whose efficacy can also be expressed probabilistically. In particular, we are interested in intermediate measures such as wearing masks, gargling, and hand washing that offer partial protection against infection while costing less than vaccination. Some kinds of vaccination rather work to reduce the contagious probability, rather than offering a perfect immunity with a certain probability. Such a mechanism including intermediate measures is called "defense against contagion" in the following text. Some recent studies[15,16,17] have proposed representing a measure of defense against contagion by reducing the risk of infection (denoted as η). Iwamura et al.[16] assumed a lower infection risk in their spatial version of the vaccination game, which they implemented by introducing a reduced infection rate per day per person into the SIR dynamics on an underlying network by means of the Gillespie algorithm.[18] Those two concepts, shown schematically in Fig. 4.3, seem analogous in the sense that one can avoid an infection stochastically, but they definitely differ in how they actually work.

We take the SIR model[19] as the baseline and extend it to consider either SIR/V or SVIR dynamics including vaccinators as SIR variants. We modify the SIR model to reproduce two different scenarios, namely imperfect vaccination (hereinafter the

[13]Kuga and Tanimoto (2018).

[14]Wu et al. (2011).

[15]Iwamura et al. (2016).

[16]Bai (2016).

[17]Cardillo et al. (2013).

[18]Gillespie (1977).

[19]Kermack and McKendrick (1927).

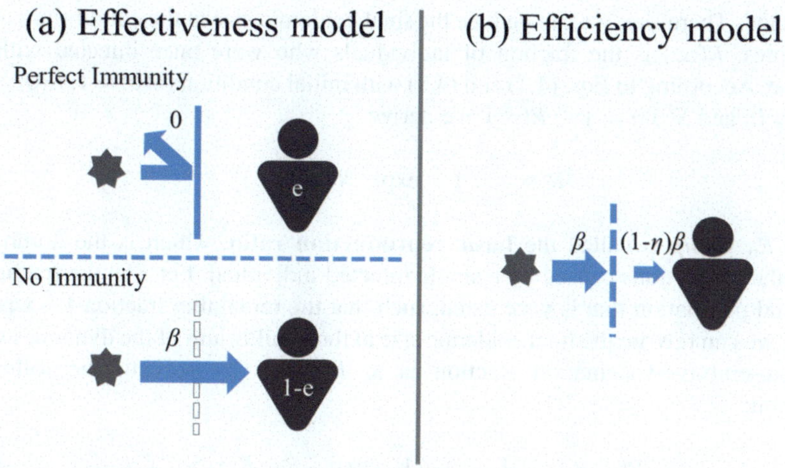

Fig. 4.3 Schematic of two concepts for avoiding infection: (**a**) effectiveness model; (**b**) efficiency model

effectiveness model) and intermediate-defense measures (hereinafter the efficiency model according to Ref [13]). We also assume a population that is infinite and ideally well mixed.

Let us presume that immunity, once acquired, works perfectly. The population is divided into three groups: susceptible individuals (S), who are currently healthy but may or may not be infected with the disease, infectious individuals (I), who are currently infected and will recover, and recovered individuals (R), who are never infected again. Immunity is acquired by either recovering from the disease or by pre-emptive vaccination. The immunity is presumed to be effective over an individual's life span. The SIR model is expressed as

$$\begin{cases} \dfrac{dS(t)}{dt} = -\beta \cdot S(t) \cdot I(t), \\ \dfrac{dI(t)}{dt} = \beta \cdot S(t) \cdot I(t) - \gamma \cdot I(t), \\ \dfrac{dR(t)}{dt} = \gamma \cdot I(t), \end{cases} \tag{4.1}$$

and,

$$S(t) + I(t) + R(t) = 1, \tag{4.2}$$

where β and γ indicate the disease-transmission rate per capita and the rate of recovery, respectively. Obviously, the SIR process always takes place in the unilateral direction, S → I → R, unlike the SIS model[20] wherein immunization efficacy is

[20]Hethcote and van den Driessche (1995).

neglected. Therefore, we can deduce the final epidemic size at the equilibrium of the dynamics. $R(\infty)$ is the fraction of individuals who were once infected with the disease. According to Eqs. (4.1) and (4.2) with initial conditions $S(0) \approx 1, R(0) = 0, I(\infty) = 0$, and $S(\infty) = 1 - R(\infty)$, we derive

$$R(\infty) = 1 - \exp[-R_0 \cdot R(\infty)]. \tag{4.3}$$

Here, $R_0 \equiv \beta/\gamma$ is called the **basic reproduction ratio**, which is the number of secondary infections caused by a single infected individual. Let x be the fraction of the total population that is vaccinated, such that the remaining fraction $1 - x$ is not. Then, we can rewrite the final epidemic size at the equilibrium of the dynamics when the pre-emptive-vaccination fraction is x, $R(x, \infty)$, by solving the following equation:

$$R(x, \infty) = (1 - x) \cdot (1 - \exp[-R_0 \cdot R(x, \infty)]). \tag{4.4}$$

This equation is obviously nonlinear and transcendental.

4.1.1.1 Effectiveness Model

A vaccinated population is separated into two classes: immune individuals obtaining perfect immunity and non-immune ones who fail to obtain immunity. Let the effectiveness of the vaccine and the vaccination coverage be e $(0 \leq e \leq 1)$ and x, respectively. The fraction of vaccinated individuals with immunity must be ex, while the fraction of non-immune individuals is $(1 - ex)$. We can express the final epidemic size R in relation to both x and time t at equilibrium $(t = \infty)$ as

$$R(x, \infty) = (1 - ex)(1 - \exp[-R_0 R(x, \infty)]). \tag{4.5}$$

$R(x, \infty)$ gives the respective fractions of four different types of individuals depending on whether they are vaccinated or non-vaccinated and whether they are healthy or infected, as summarized in Table 4.1.

4.1.1.2 Efficiency Model

Let the efficiency of an intermediate defense measure for avoiding infection be η $(0 \leq \eta \leq 1)$, describing how the defense measure can decrease the probability of being infected. In the following formulation of the efficiency model, we temporarily regard the vaccinated state as the state prepared with an intermediate defense measure, making it convenient to compare it with the aforementioned effectiveness model. We describe the epidemic-spreading dynamics using the compartment model whereby individuals in a population can be classified into susceptible (S), infected

Table 4.1 Fractions of four types of individual using the effectiveness model

Strategy/state	Healthy	Infected
Vaccinated	$x(e + (1 - e)\exp[-R_0R(x,\infty)])$	$x(1 - e)(1 - \exp[-R_0R(x,\infty)])$
Non-vaccinated	$(1 - x)\exp[-R_0R(x,\infty)]$	$(1 - x)(1 - \exp[-R_0R(x,\infty)])$

(I), recovered (R), and vaccinated (V) states. A non-vaccinated (more precisely, non-prepared with intermediate defense measures) susceptible individual may become infected if they are exposed to infectious individuals at the disease-transmission rate β [per day per person]. An S individual prepared with intermediate defense measures may also become infectious at a rate $(1 - \eta)\beta$. An infected individual recovers at the recovery rate γ [per day].

The SVIR model we use to describe such a condition is

$$\begin{cases} \dfrac{dS(x,t)}{dt} = -\beta S(x,t)I(x,t), \\ \dfrac{dV(x,t)}{dt} = -(1-\eta)\beta V(x,t)I(x,t), \\ \dfrac{dI(x,t)}{dt} = \beta S(x,t)I(x,t) + (1-\eta)\beta V(x,t)I(x,t) - \gamma I(x,t), \\ \dfrac{dR(x,t)}{dt} = \gamma I(x,t), \end{cases} \tag{4.6}$$

with the presumed set of initial values $S(x,0) = 1 - x$, $V(x,0) = x$, and $I(x,0) = 0$. The following constraint is requisite:

$$S(x,t) + V(x,t) + I(x,t) + R(x,t) = 1. \tag{4.7}$$

Because the population is not completely susceptible, it is accurate to use a control reproduction number, R_c, instead of the basic reproduction number, R_0. In this case, R_c can be estimated as

$$R_c = \frac{\beta}{\gamma}[S(x,0) + (1-\eta)V(x,0)] = R_0[S(x,0) + (1-\eta)V(x,0)]. \tag{4.8}$$

The final epidemic size and other fractions can be expressed as;

$$S(x,\infty) = (1-x)\exp[-R_0R(x,\infty)], \tag{4.9}$$

$$V(x,\infty) = x\exp[-(1-\eta)R_0R(x,\infty)], \tag{4.10}$$

$$R(x,\infty) = 1 - (1-x)\exp[-R_0R(x,\infty)] - x\exp[-(1-\eta)R_0R(x,\infty)]. \tag{4.11}$$

In the limit of this process, the respective fractions of the four different types of individual at equilibrium are as summarized in Table 4.2.

Table 4.2 Fractions of four types of individual using the efficiency model

Strategy/state	Healthy	Infected
Vaccinated	$x \exp[-(1-\eta)R_0R(x,\infty)]$	$x(1- \exp[-(1-\eta)R_0R(x,\infty)])$
Non-vaccinate	$(1-x)\exp[-R_0R(x,\infty)]$	$(1-x)(1- \exp[-R_0R(x,\infty)])$

Comparing Tables 4.1 and 4.2, it is worth noting that the "success probability of free-riding" is always given by $\exp[-R_0R(x,\infty)]$, regardless of whether we presume perfect or imperfect vaccination, or even intermediate defense measures.

Figure 4.4 shows the final epidemic size (FES) for different levels of vaccination coverage using the effectiveness and efficiency models. From Fig. 4.4, the so-called critical-vaccination coverage that eradicates an epidemic spread can be read from the border of the extinct phase at which FES = 0. This border suggests the critical vaccination coverage for suppressing the spread of an infection, which can be determined analytically as $x_c = (1 - 1/R_0)/(1 - \varepsilon)$ for the efficiency model and $x_c = (1 - 1/R_0)/e$ for the effectiveness model. Clearly, as long as a less-reliable defense measure is provided, say $\eta < 0.6$, we cannot prevent an epidemic from breaking out, even if all individuals use that particular defense measure.

4.1.1.3 Relationship Between Effectiveness and Efficiency Models

Let us establish an explicit relationship between the effectiveness and efficiency models. Eqs. (4.3) and (4.11) give

$$\begin{aligned} R(x,\infty) &= (1-ex)(1-\exp[-R_0R(x,\infty)]) \\ &= 1-(1-x)\exp[-R_0R(x,\infty)] - x\exp[-(1-\eta)R_0R(x,\infty)]. \end{aligned} \tag{4.12}$$

Equation (4.8) gives the relationship between e and η as

$$e = \frac{\exp[-(1-\eta)R_0R(x,\infty)] - \exp[-R_0R(x,\infty)]}{1-\exp[-R_0R(x,\infty)]}. \tag{4.13}$$

Figure 4.5 shows the relationship between e and η indirectly. The vacant region for $x \approx 1$ and $\eta \approx 1$ is due to the fact that a multivalued e–η relationship is inevitable. It is worth noting that, for smaller values of η, e appears insensitive to x, implying that e constantly relates to η irrespective of x. By contrast, for larger η (i.e., $\eta > 0.6$), the colored contours appear slanted, suggesting that the e–η relationship becomes fully dependent on x. In other words, epidemic dynamics that assume a higher effectiveness of vaccination (close to perfect immunity) work differently from those that assume an equivalent efficiency of the intermediate defense measure, which is influenced strongly by how many individuals use that particular vaccination or defense measure.

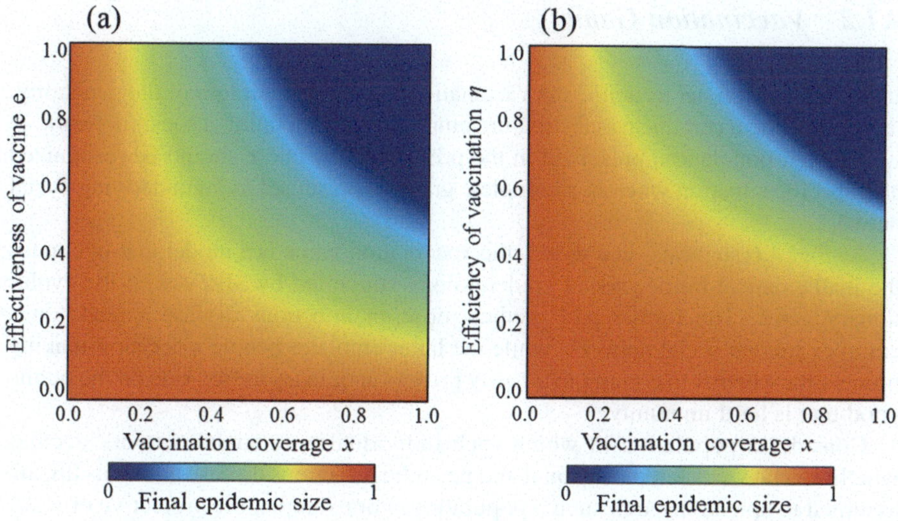

Fig. 4.4 Final epidemic size according to vaccination coverage and (**a**) effectiveness *e* (effectiveness of a vaccination) or (**b**) efficiency η (efficiency of an intermediate defense measure). We presume that $R_0 = 2.5$, which is applied consistently in this study

Fig. 4.5 Effectiveness *e* plotted on (x, η) plane

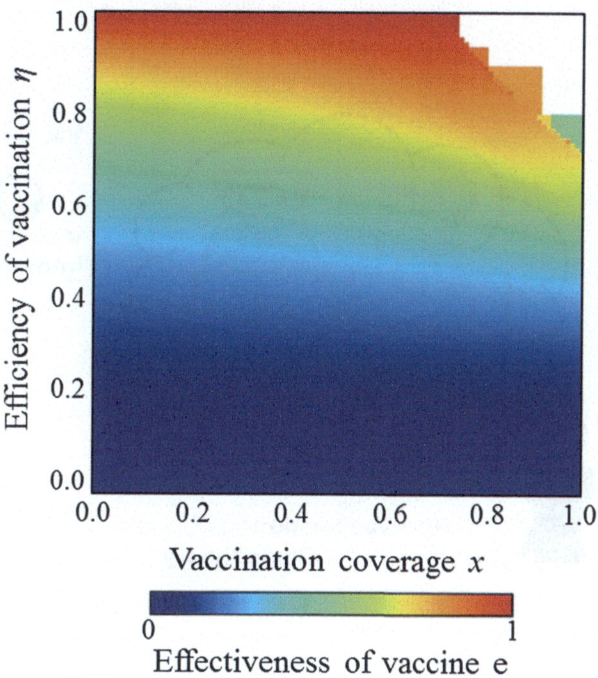

4.1.2 Vaccination Game

In this subsection, let us define the vaccination game. Without loss of the generality, let us presume a certain spatial structure among agents, defining a social network. A well-mixed population, presumed in the previous subsection, should be recognized as a specific situation wherein a complete graph is presumed as an underlying social network.

As Fig. 4.6 suggests, in a world, the vaccination game is a model that integrates the mathematical framework of epidemiology supported by SIR/V with the evolutionary game. The former part predicts how an infectious disease spreads on a complex human social network, while the latter emulates people's decision-making process for whether to commit to a costly provision or to try to free-ride on the public good that is herd immunity.

Consider a population in which each individual on a social network decides whether to be vaccinated. Seasonal and periodic infectious diseases, such as flu, are assumed to spread through such a population. For example, the protective efficacy of a flu vaccine persists for less than a year because of waning antibodies and year-to-year changes in the circulating virus. Therefore, under a voluntary vaccination program, individuals must decide every year whether to be vaccinated. Thus, the dynamics of our model consists of two stages: the first stage is a vaccination campaign and the second is an epidemic season.

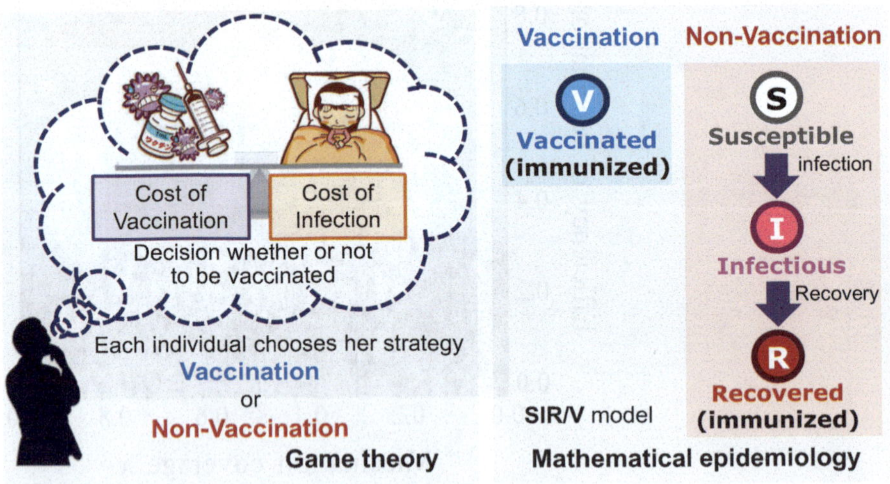

Fig. 4.6 Vaccination game

4.1.2.1 The First Stage: The Vaccination Campaign

Here, in this stage, each individual makes a decision whether to get vaccinated before the beginning of the seasonal epidemic, i.e., before any individuals are exposed to the epidemic strain. Vaccination imposes a cost, C_v, on each individual who decides to be vaccinated. The cost of vaccination includes the monetary cost and other perceived risks, such as adverse side effects. If an individual is infected, he incurs the cost, C_i, of infection.

In real, as we discussed already, the vaccination does not provide perfect immunity. Individuals who unfortunately are infected despite taking either the vaccination or the intermediate defense measure against contagion are assigned the cost $C_v + C_i$ of vaccination and infection. Needless to say, an individual neither vaccinated nor taking any intermediate defense measure against contagion faces the risk of being exposed to infection during a season.

To simplify the evaluation of each individual's payoff, without loss of generality, we rescale the cost by defining a relative cost of vaccination, namely $C_r = C_v/C_i$ ($0 \leq C_r \leq 1$; $C_i = 1$). Consequently, the payoff of each individual at the end of an epidemic season depends on his/her final state. Table 4.3 summarizes the payoff whether committing to a provision (either vaccination or intermediate defense measure against contagion) or not and whether having been healthy or infected.

4.1.2.2 The Second Stage: The Epidemic Season

Here, at the beginning of this stage, the epidemic strain enters the population, and a number I_0 (sufficiently small compared with the total population) of randomly selected susceptible individuals are identified as the initially infected ones. Then, the epidemic spreads according to SIR/V dynamics. At the end of one epidemic season, we can observe the final epidemic size, previously discussed in Eq. (4.4).

It is likely that, after one epidemic season, an individual would reevaluate their decision of whether or not to commit to a provision based on whether or not they were infected during the season. For instance, they may shift to non-vaccination if quite a few of their neighbors successfully avoided infection during the season, or if media claims that the current infection rate is not as serious as expected. Depending on how each individual adjusts their strategy, as defined in the later sub-section, the fractions committing or not committing to a provision in the next season evolve. Although what we seek is not the annual dynamics in terms of the final epidemic size and vaccination coverage over long epidemic seasons, we repeat the set of one vaccination campaign and one epidemic season many times until its time evolution

Table 4.3 Payoff structure determined at the end of an epidemic season

Strategy/state	Healthy	Infected
Vaccinated	$-C_r$	$-C_r - 1$
Non-vaccinated	0	-1

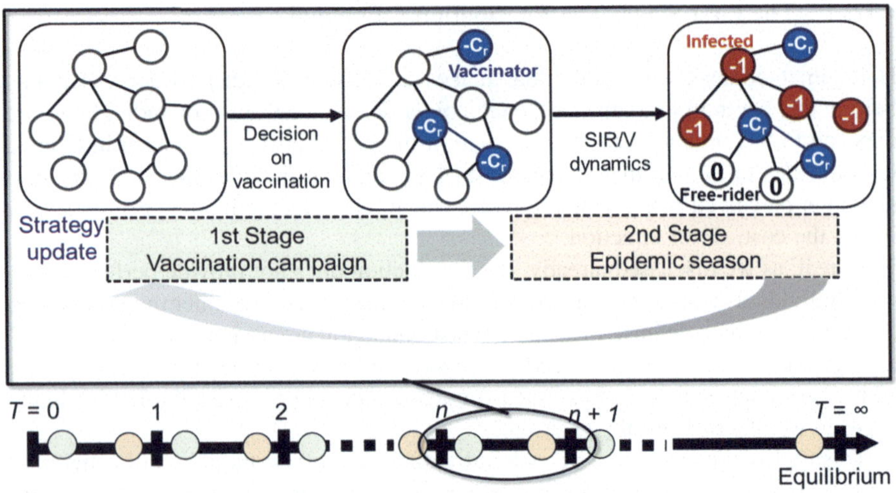

Fig. 4.7 Time sequence in a vaccination game. T denotes the number of epidemic seasons

reaches a certain equilibrium. We then evaluate the social equilibrium where the final epidemic size, vaccination coverage, and social-average payoff are measured. The vaccination-game concept is explained in Fig. 4.7.

4.1.3 Multiagent Simulation (MAS) Approach

The classic SIR or SIR/V model, discussed in Sect. 4.1.1, is given by coupled (integro-) differential equations and does not assume any spatial structure for the population. Such an analytical approach can be applied to cases presuming spatial structure amid individuals. However, a more convenient and powerful multiagent simulation (MAS) approach can be applied to various cases presuming a spatial structure of finite population size.

Suppose that the whole population has a spatial structure, represented by a network consisting of nodes and links. The dynamics of SIR/V on a spatially structured population are not captured by a system of differential equations; thus, we numerically simulate an epidemic spreading on a network using the Gillespie algorithm[18] in the extended SIR/V model. For social networks, we can account for any topology in the MAS model. Many interesting toplogies have been discussed; rings (representing 1D regular graphs), lattices (representing 2D regular graphs), random regular networks (RRGs), Barabási-Albert scale-free (BA-SF) networks,[21]

[21]Barabási and Albert (1999).

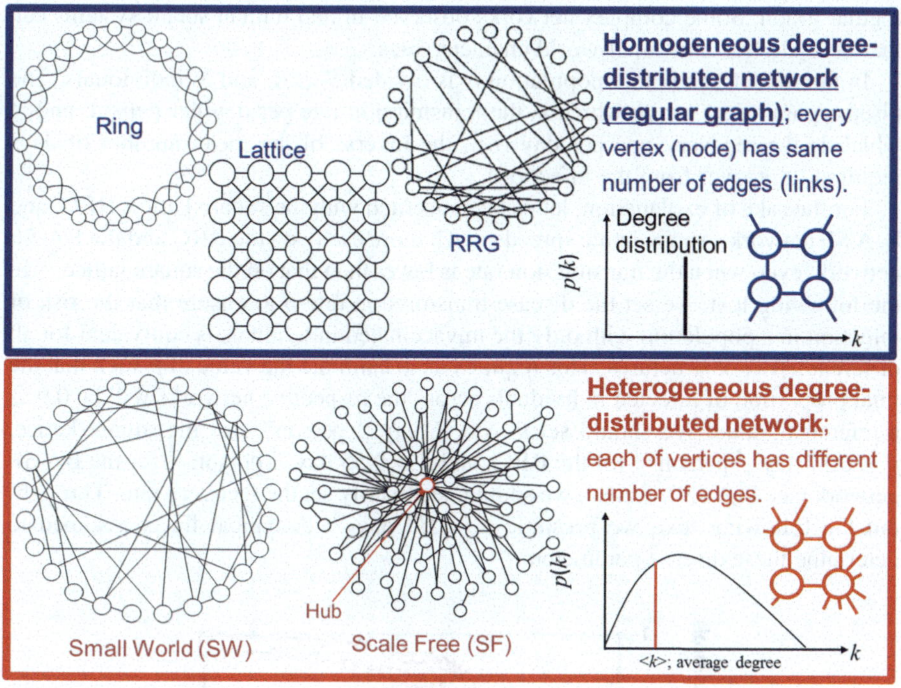

Fig. 4.8 Representative topologies

Erdős–Rényi random (E-R random) graphs,[22] and small-world (SW) network.[23] Figure 4.8 gives some such models. Meanwhile, a couple of network properties have been defined to evaluate its topology and complexity; such as average degree, $<k>$, degree distribution, $P(k)$, average path length, average cluster coefficient, and assortative coefficient. In rings, lattices, and RRGs, each vertex has the same degree k, i.e., the same number of links, and so these are called homogenous degree-distributed networks. The degree distribution of a scale-free graph obeys a scale-free distribution, and that of the E-R random graph obeys a Poisson distribution. These are classified as heterogeneous degree-distributed networks. A scale-free network has a small number of agents (called "hubs") with a huge number of links, while the vast majority of agents have a small number of links. It is why its degree-distribution is scale-free. And it is the substance of scale-free network. A small-world network can be constructed from a regular graph such as a ring or a lattice. Starting from such a regular graph, severing links with "short-cut" probabilities (usually presumed to be a small value), and re-connecting them randomly, the graph would become a small-world. One of the most important characteristics is that an SW network has a quite small average path length compared with the original

[22]Bollobás (1985).

[23]Watts and Strogatz (1998).

regular graph. Some complex networks observed in real human social systems can feature scale-free and small-world characteristics.

In the model, the whole population N is divided S, I, R, and V individuals. The disease parameters are β, which is the transmission rate per day per person, and γ, which is the recovery rate per day (i.e., the inverse of the mean number of days required to recover from the infection).

For the sake of explanation, let us be concerned with the square lattice, RRG, and B-A SF network. An epidemic spreads much more easily on the RRG and the BA-SF network, even when the transmission rate is lower than that on the square lattice.[24] In the following text, we set the disease-transmission rate β to ensure that the risk of infection in a population with only the unvaccinated individuals is equivalent for all different network structures. This requires us to calibrate the value of β such that the final proportion of infected individuals across the respective networks will be 0.9 as a reference value. We should set $\beta = 0.46$ day^{-1} person^{-1} for the square lattice, $\beta = 0.37$ day^{-1} person^{-1} for the RRG, and $\beta = 0.55$ day^{-1} person^{-1} for the BA-SF network (see Fig. 4.9).[25] Also, we should necessarily set the recovery rate. Throughout the following text, we presume $\gamma = 1/3$ day^{-1}. A typical flu is assumed to determine these disease parameters.

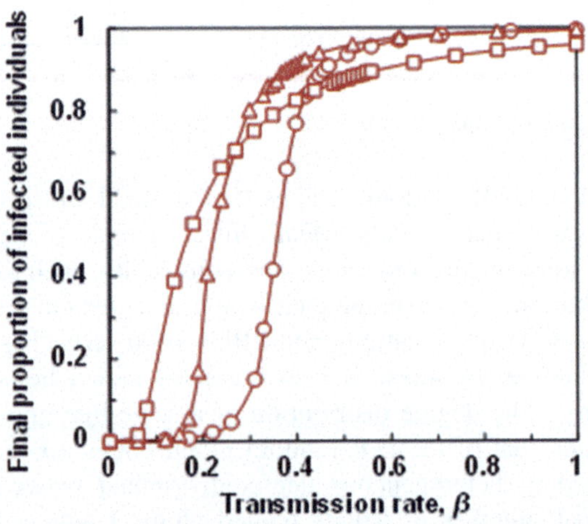

Fig. 4.9 Final proportion of infected individuals as a function of transmission rate β when no individuals are vaccinated on each network: square lattice (circles), random regular graph (RRG) (triangles), Barabási-Albert scale-free (BA-SF) network (squares). For the lattice (circles): population size $N = 70 \times 70$ with von a Neumann neighborhood, recovery rate $\gamma = 1/3$ day^{-1}, seeds of epidemic spread $I_0 = 5$. For RRG (triangles): population size $N = 4900$, degree $k = 4$, recovery rate $\gamma = 1/3$ day^{-1}, seeds of epidemic spread $I_0 = 5$. For BA-SF network (squares): population size $N = 4900$, average degree $\langle k \rangle = 4$, recovery rate $\gamma = 1/3$ day^{-1}, seeds of epidemic spread $I_0 = 5$. Each plotted point represents an average over 100 runs

[24]Keeling and Eames (2005), Pastor-Satorras and Vespignani (2001).

[25]Fukuda et al. (2014).

At the beginning of each epidemic season, we randomly place a number I_0 of initially infected agents. According to the Gillespie algorithm,[19] we exactly follow the epidemiological dynamics of the SIR/V model. This is analogous to a simulation based on the percolation theory.[26] The epidemic season lasts until no infection exists in the population. Each individual who gets infected during the epidemic incurs the cost of infection, C_i. However, the cost paid by a "free-rider" who does not vaccinate and still is free from infection is zero.

As shown in Fig. 4.7, right after one epidemic season, another vaccination campaign begins, during which each agent in the network refreshes their strategy. We repeat the set of one vaccination campaign and one epidemic season many times until its time evolution reaches a certain equilibrium to obtain the final epidemic size. The vaccination coverage and the social average payoff are measured.

4.1.4 Decision-Making Process Concerning Vaccination

In this sub-section, we describe how each agent adapts their strategy, regardless of whether they commit to a provision in a vaccination campaign; this is the first stage in the vaccination game, schematically shown in Fig. 4.7. One's motivation in adjusting their strategy over time is to maximize their own payoff as long as each individual is presumed to behave in a rational way. According to what has been suggested by evolutionary game theory, even if one behaves in an altruistic manner, there must be compensation in the long run. If not, such cooperative behavior cannot be evolutionary stable. Roughly speaking, one can update their strategy through social imitation or self-estimation.

Social imitation is "copying" from others. The information source, from which a focal agent copies, might be an acquaintance, whether a neighbor, relative, friend, or someone sharing common benefits with the focal agent; or they might be on the media. In the latter case, the information is global, unlike in the former case relying on local information.

The second idea, self-estimation, differs from copying. In many real situations concerning decision making, it is conceivable that an agent will engage not in social imitation but drawing a personality-independent decision that directly suggests that one of the alternative strategies to be taken in the next time-step is conceivable. In this case, a focal agent quantitatively estimates the given situation based on the observation of what currently happens to his neighbors including himself. That quantitative estimation teaches him which strategy is stochastically best for the next time-step. One commonly shared idea concerning the strategy-update rule in evolutionary game theory, the "aspiration model",[27] belongs to this second idea.

---

[26]Sahimi (1994).

[27]There have been many previous studies on the aspirational model; let us suggest some representative literature below; Macy and Flache (2002), Chen and Wang (2008).

4.1.4.1 Social Imitation

As mentioned above, this concept comes from how one copies others. We note two ideas.

Individual-Based Risk Assessment (IB-RA)
Fu et al. pioneered a model[10] in which agents, spatially distributed on an underlying network and exposed to infectious risk, learn whether or not to vaccinate from one of their neighbors. This idea exactly reflects the assumption of the vast majority of studies dealing with spatial prisoner's dilemma (SPD) games, namely pairwise comparison based on a Fermi function (as described by PW-Fermi). Agent i randomly selects agent j from his neighbors. Let us assume that their payoffs are π_i and π_j, respectively. The probability of agent i copying agent j's strategy, s_j, either vaccination or non-vaccination, instead of his own strategy, s_i, is $P\left(s_i \leftarrow s_j\right)$, which is defined as

$$P\left(s_i \leftarrow s_j\right) = \frac{1}{1+\exp\left[\frac{\pi_i-\pi_j}{\kappa}\right]}, \tag{4.14}$$

where κ indicates the sensitivity to the gain difference. For $\kappa \to \infty$ (weak-selection pressure), an individual i is insensitive to the payoff difference $\pi_i - \pi_j$ against another individual j and the probability $P(s_i \leftarrow s_j)$ approaches 1/2 asymptotically, regardless of the payoff difference. For $\kappa \to 0$ (strong selection pressure), individuals are sensitive to the payoff difference, and they definitely copy the successful strategy that earns the higher payoff, even if the difference in the payoff is very small. We assume that $\kappa = 0.1$.

Strategy-Based Risk Assessment (SB-RA)[28]
Equation (4.14) indicates that as the negative-payoff difference increases, the probability that an individual will change their strategy to that of their successful neighbor increases. Observing Eq. (4.14) from a different viewpoint, this rule of strategy adaptation can be interpreted as follows: each individual evaluates both the risk of maintaining their own strategy and imitating that of their opponent and then selects the one with the smaller risk. In this method, each individual i assesses the risk based only on one certain individual j because Eq. (4.14) uses only the payoff of i's opponent (individual j). Thus, we call the updating rule, described by Eq. (4.14) an individual-based risk-assessment updating rule (IB-RA).

However, when we assume that the information regarding the consequences of adopting a certain strategy are disclosed to the society and everyone in the population has access to this information, then individuals no longer rely heavily on the payoff of any one neighbor. Instead, in adapting their strategy, they tend to assess the risk based on a socially averaged payoff that results from adopting a certain strategy.

To reflect the above situation, we propose the following modified imitation probability:

$$P(s_i \leftarrow s_j) = \frac{1}{1 + \exp\left[\frac{\pi_i - <\pi_{s_j}>}{\kappa}\right]}, \tag{4.15}$$

where $< \pi_{s_j} >$ is an average payoff obtained from a collective payoff over individuals who adopt the same strategy as that of a randomly selected neighbor j of the individual i. The sampling number is a control parameter that ranges from only one individual (i.e., only one of i's neighbors, j) to all individuals among the whole population who adopt the same strategy as j. That is, if s_j is the strategy of vaccination (cooperation, C), then $< \pi_{s_j} >= -C_r$ (since the payoff of a vaccinated individual is uniquely determined[28]); whereas, if s_j is the strategy of no-vaccination (defection, D), then $< \pi_{s_j} >$ takes a value between 0 and 1, depending on the fractions of infected and healthy individuals (free-riders) with the strategy s_j in the population at the end of the epidemic. Moreover, if sampling is impossible because the population size of individuals with the strategy s_j is too small, the individual i uses the payoff of one randomly selected neighbor instead of $< \pi_{s_j} >$ in Eq. (4.15), leading to an expression that is the same as Eq. (4.14). Thus, when the sampling rate is set to zero, Eq. (4.15) reduces to Eq. (4.14).

Equation (4.15) implies that an individual i assesses the risk of changing their strategy based on the payoff attained by adopting a certain strategy, and not the payoff attained by a certain other individual. Thus, we call the updating rule (4.15) a "strategy-based risk assessment updating rule" (SB-RA). Note that, risk assessment based on the consequences of a vaccination strategy is the same as that based on a unique individual because the immune effect of vaccination is perfect during an epidemic season. However, for the no-vaccination strategy, the risk may differ from season to season because the degree of the epidemic may differ.

Figure 4.10 schematically summarizes IB-RA and SB-RA.

4.1.4.2 Self-Estimation[15]

Instead of "copying probability from one of the neighbors," we directly assign an agent a "probability of vaccinating," P_C, triggered by his consciousness of how dangerous it is to ignore vaccination. Namely,

$$P_C = \frac{1}{1 + \exp[(C_r - < C_D >)/\kappa]}, \tag{4.16}$$

[28]This is true only when a vaccination brings perfect immunity. When we presume a general provision, we have either imperfect vaccination dealt by the effectiveness model or an intermediate protection measure for contagion by the efficiency model (see Fig. 4.3). The cooperator's payoff is stochastically variable. In this case, we must evaluate $< \pi_{s_j} >$ fairly, as in the defector's case.

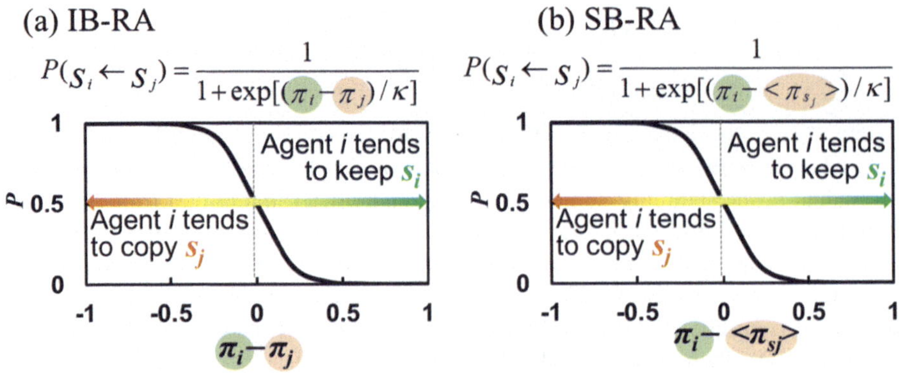

Fig. 4.10 Strategy update based on the concept of social imitation; (**a**) IB-RA, (**b**) SB-RA

$$\frac{< C_D >= C_i \cdot n_i +^i C_f \cdot n_f}{n_D,} \tag{4.17}$$

where $C_i (= 1)$is the cost of being infected, C_f is the cost of free riding, which is zero, and n_D, n_i, and n_f are the numbers of non-vaccinators, infected agents, and free riders, respectively, in the agent's neighborhood. Therefore, $n_D = n_i + n_f$. $<C_D>$ indicates the average payoff of non-vaccinators in agent *i*'s neighborhood. In the following text, let us call this model "direct commitment (DC)".

One problem that arises is how to evaluate Eq. (4.17) if there are no non-vaccinators in agent *i*'s neighborhood. We establish the following four cases as our sub-model.

Case 1: Agent *i* retains their strategy.
Case 2: As a substitute, we assume that $P_C = 1\text{-}C_r$.
Case 3: Agent *i* switches to the strategy opposite theirs.
Case 4: Substituting $<C_D> = 0$, we continue to rely on Eq. (4.16).

Case 1 expresses the fact that people tend to maintain the status quo. Case 2 assumes that an agent relies on the vaccination cost as alternative information. Case 3 assumes that an agent tends to take an inverse strategy if they are stalemated due to lack of information. Case 4 assumes that an agent behaves in an optimistic manner by assuming that free riders can be successful.

In the following text, we show the result of comparing Cases 1 and 4 of the DC model by means of MAS.

We assume that $N = 4900$ and $I_0 = 5$. We also assume that $\beta = 0.46$ and $\gamma = 1/3$ (the flu is assumed). In a simulation episode, one time-step consisting of the first stage, a vaccination campaign and the second stage, and an epidemic season continues until 3000 time-steps have passed. In a simulation study with varying relative vaccination cost, C_r, we observe the average vaccination coverage and

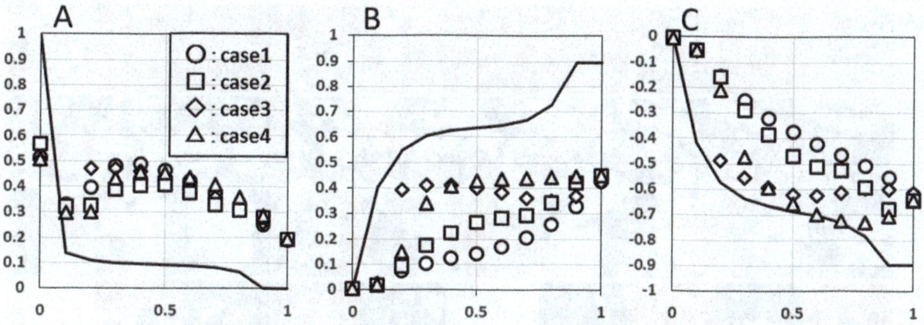

Fig. 4.11 Relationship between the vaccination cost and (**a**) the vaccination coverage, (**b**) the final epidemic size, and (**c**) the average social payoff for each of the four cases. Different symbols indicate the four cases, while the solid line indicates the result of the default setting according to Fu et al. (2011)

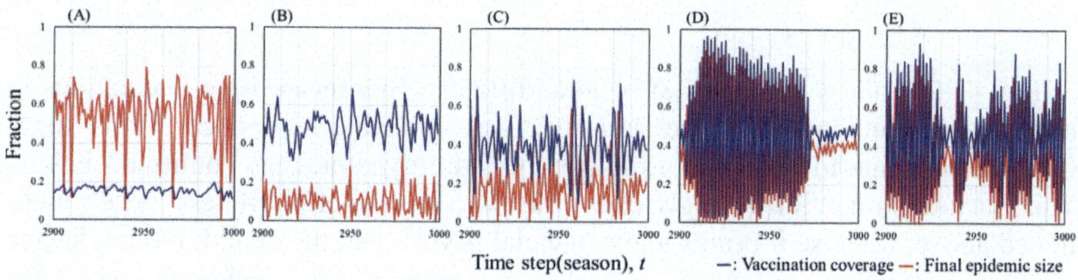

Fig. 4.12 Time evolution of the vaccination coverage (blue) and the final epidemic size (red) assuming $C_r = 0.3$: (**a**) default case, (**b**) Case 1, (**c**) Case 2, (**d**) Case 3, and (**e**) Case 4

final epidemic size in the last 1000 time-steps. The statistics shown below are based on 100 independent simulation episodes.

In Fig. 4.11, we show (a) the vaccination coverage, (b) the final epidemic size, and (c) the average social payoff versus the vaccination cost, C_r. Except at $C_r = 0$, our new adaptation model shows higher vaccination coverage, therefore leading to smaller final epidemic sizes than seen in the default model. However, note that, as far as the social payoff is concerned, Cases 3 and 4 show worse performance than either Cases 1 or 2 for the range of reasonable vaccination cosst, although they seem better than the default case.

To further examine the results, Fig. 4.12 shows a typical time evolution of 100 realizations for the last 100 time-steps prior to quasi-equilibrium in each case (Fig. 4.12b–e) with the default model (Fig. 4.12a) assuming $C_r = 0.3$. In the default case, as a general tendency, we see that larger final epidemic sizes with small time-fluctuations (compare Case 1 (Fig. 4.12b) and Case 2 (Fig. 4.12c)) result from lower but more stable vaccination coverage. Conversely, in Case 2 and more clearly in Case 1, relatively higher and stable vaccination coverage successfully results in stably lower final epidemic sizes. Interestingly, the situation we observe in Cases

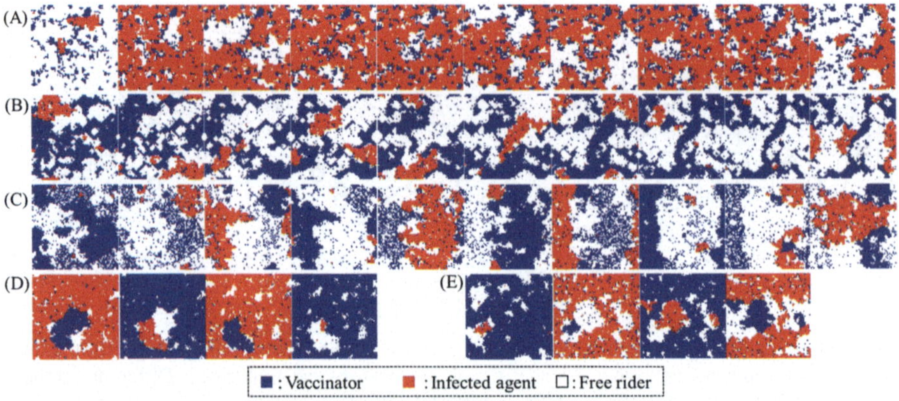

Fig. 4.13 Snapshots after 2940 time-steps assuming $C_r = 0.3$: (**a**) Default case, (**b**) Case 1, (**c**) Case 2, (**d**) Case 3, and (**e**) Case 4

3 (Fig. 4.12d) and 4 (Fig. 4.12e) is quite different. Significant time-fluctuations in both the vaccination coverage and the final epidemic size occur. This time-fluctuation seems to have two time-step-periodic dynamics, as confirmed below. This fluctuation may unwillingly cause the "vaccination-effectiveness" result mentioned above, because it brings a lower social payoff than the default model despite higher vaccination coverage and lower average final epidemic size for $0.3 \leq C_r \leq 0.6$.

Figure 4.13 offers further insight by showing continuous snapshots after 2940 time-steps for each of the four cases. Obviously, Cases 3 (Fig. 4.13d) and 4 (Fig. 4.13e) show two-time-step-periodic flipping in which a situation with vaccinators with a small number of infected agents who failed to free ride follows a situation with infected agents with a small number of vaccinators. This inevitably results in a pandemic-like situation in Cases 1 (Fig. 4.13b) and 2 (Fig. 4.13c) every other time-step, because the majority of vaccinators form small vaccinator clusters and are less spatially spread-out over the entire domain.

One plausible reason why these settings, especially those of Case 4, bring about such acute two-time-step flipping can be formulated as follows. From the inherent nature of its definition, Case 4 sees the non-vaccination strategy as being more advantageous than the vaccination strategy if there are no infected neighbors. Our model framework in not only Case 4 but also in other cases urges infected agents to adopt a vaccination strategy in the next time-step. This implies that agents in Case 4 tend to adopt the defective strategy (non-vaccination) if the outcome of the current time-step is good, whereas they adopt a cooperative strategy (vaccination) if the outcome is bad. This specific feature consequently results in the time flipping that was also observed in Win-Stay and Lose-Shift (WSLS)[29] of PD games and is a

[29]Imhof et al. (2007).

typical self-reflecting strategy (taking either the same offer to the current one or its opposite), unlike the copying-from-others' strategy seen in tit-for-tat games. As confirmed above, Cases 3 and 4 contain some fragments of a "self-reflecting strategy." One recent study[30] reports that some strategic adaptations based on the concept of WSLS can enhance cooperation for spatial-PD games. This is because sparsely located cooperative agents showing time-flipping manners with time-alternating defection are able to realize a reasonable amount of mutual reciprocity. This is interesting because the mechanism appears to be very different from the usual network reciprocity as previously understood, in which a situation of compactly clustered cooperators would be more likely to result in efficient network reciprocity. However, in the vaccination game, which has a different game structure than PD games, Cases 3 and 4 somehow contain a "self-reflecting" feature that does not result in any preferable consequences.

Unlike Cases 3 and 4, Cases 1 and 2 are able to achieve a preferable consequence, whereby a higher social payoff is established than that in the case of the default model. Figure 4.13b, c indicate that, in both cases, quite a few vaccinators are sparsely located and, consequently, vaccinators can ubiquitously exist in any corner of the domain. This is crucially important for suppressing the spread of epidemics.[31]

As discussed, we have established a new strategy-adaptation idea in the vaccination game. Our update rule does not provide a "copying probability" from a focal agent's neighbor, as in conventional models; rather, it directly gives a "vaccination probability" derived from a stochastic comparison between the vaccination cost commonly disclosed in public and the expected benefit resulting from adopting the non-vaccination strategy observed in an agent's neighborhood. We further define four subordinate models depending on how an agent behaves if they do not encounter a non-vaccinator among their neighbors.

The simulation results show that our new adaptation model generally results in higher vaccination coverage and smaller final epidemic sizes than those in the conventional model, which assumes social imitation of one of the neighbors.

However, depending on the subordinate models, there were two final consequences that either efficiently suppress epidemic spreading or do not. Specifically, the case assuming that an agent that retains their strategy even if there are no neighboring defectors (non-vaccinators) (Case 1) allows vaccinators to be sparsely located in the domain, successfully hampering the spread of an epidemic in this domain. Conversely, the case assuming that an agent takes the reverse strategy if there are no defectors (Case 3) or assumes that a free ride will be successful if they have no defectors among their neighbors (Case 4) results in an acute time-flipping behavior, which allows huge pandemics every two time-steps.

---

[30] Amaral et al. (2016).

[31] Fukuda et al. (2015).

4.1.4.3 Mean-Field Approximation (MFA) for the Three Updating Rules; IB-RA, SB-RA, and DC

In Sect. 4.1.1, we discussed the analytic model of SIR/V presuming an infinite and well-mixed population. Here, no spatial structure—i.e., no network that connects all of the individuals—is considered. Hence, when we apply the vaccination game to the analytic model, the strategy-update rule should be introduced. Since any spatial structure is considered, we must rely on the so-called mean-field approximation to evaluate a neighbor's payoff.

Individual-Based Risk Assessment (IB-RA)

In the present framework, there are four classes of individual in relation to cost burden: (i) a successful free-rider (SFR) who pays nothing; (ii) a failed free-rider (FFR) who pays -1; (iii) an infected vaccinator (IV) who pays $-C_r - 1$; and (iv) a healthy vaccinator (HV) who pays $-C_r$. Each individual can choose from one of two strategies: vaccination (hereinafter V) or non-vaccination (hereinafter NV). Thus, the transition probability that affects the time transition of x, which should be considered in the IB-RA rule, is covered by one of the following eight cases:

$$P(HV \leftarrow SFR) = \frac{1}{1 + \exp[-(0 - (-C_r))/\kappa]}, \tag{4.18a}$$

$$P(HV \leftarrow FFR) = \frac{1}{1 + \exp[-(-1 - (-C_r))/\kappa]}, \tag{4.18b}$$

$$P(IV \leftarrow SFR) = \frac{1}{1 + \exp[-(0 - (-C_r - 1))/\kappa]}, \tag{4.18c}$$

$$P(IV \leftarrow FFR) = \frac{1}{1 + \exp[-(-1 - (-C_r - 1))/\kappa]}, \tag{4.18d}$$

$$P(SFR \leftarrow HV) = \frac{1}{1 + \exp[-(-C_r - 0)/\kappa]}, \tag{4.18e}$$

$$P(SFR \leftarrow IV) = \frac{1}{1 + \exp[-(-C_r - 1 - 0)/\kappa]}, \tag{4.18f}$$

$$P(FFR \leftarrow HV) = \frac{1}{1 + \exp[-(-C_r - (-1))/\kappa]}, \tag{4.18g}$$

$$P(FFR \leftarrow IV) = \frac{1}{1 + \exp[-(-C_r - 1 - (-1))/\kappa]}. \tag{4.18h}$$

Strategy-Based Risk Assessment (SB-RA)
As discussed above, SB-RA modifies the imitation probability to reflect the situation in which an individual tends to assess the risk based on a socially averaged payoff because of the prevalence of information about epidemics (probably given by the media). In the analytic framework, the transition probability that we must consider now is one of the following:

$$P(HV \leftarrow NV) = \frac{1}{1 + \exp[-(\langle \pi_D \rangle - (-C_r))/\kappa]}, \tag{4.19a}$$

$$P(IV \leftarrow NV) = \frac{1}{1 + \exp[-(\langle \pi_D \rangle - (-C_r - 1))/\kappa]}, \tag{4.19b}$$

$$P(SFR \leftarrow V) = \frac{1}{1 + \exp[-(\langle \pi_C \rangle - 0)/\kappa]}, \tag{4.19c}$$

$$P(FFR \leftarrow V) = \frac{1}{1 + \exp[-(\langle \pi_C \rangle - (-1))/\kappa]}. \tag{4.19d}$$

Direct Commitment (DC)
Direct commitment is the representative framework of "self-estimation", which differs from "social-imitation" methods such as IB-RA and SB-RA. Applying MFA, the transition probability that we must consider now is one of the following:

$$P(V \leftarrow NV) = \frac{1}{1 + \exp[-(\langle \pi_D \rangle - \langle \pi_C \rangle)/\kappa]}, \tag{4.20a}$$

$$P(NV \leftarrow V) = \frac{1}{1 + \exp[-(\langle \pi_C \rangle - \langle \pi_D \rangle)/\kappa]}. \tag{4.20b}$$

4.1.5 *Vaccination Game Through Analytic Approach*

Here, let us revisit the analytical approach. Recall that x means the fraction of the total population that either vaccinates or takes intermediate defensive measures. Looking back at Fig. 4.7, strategy updating takes place after each epidemic season, as defined above, thereby inevitably changing x. We have two different epidemic models, namely the effectiveness model and the efficiency model, and three different updating rules, namely IB-RA, SB-RA, and DC. Hence, we establish the following six different types of dynamics:

Effectiveness model + IB-RA:

$$\begin{aligned}\frac{dx}{dt} = \; & x(1-x)(e+(1-e)\exp[-R_0R(x,\infty)])\exp[-R_0R(x,\infty)] \\ & (P(SFR \leftarrow HV) - P(HV \leftarrow SFR)) \\ & + x(1-x)(e+(1-e)\exp[-R_0R(x,\infty)]) \\ & (1-\exp[-R_0R(x,\infty)])(P(FFR \leftarrow HV) - P(HV \leftarrow FFR)) \\ & + x(1-x)(1-e)(1-\exp[-R_0R(x,\infty)])\exp[-R_0R(x,\infty)] \\ & (P(SFR \leftarrow IV) - P(IV \leftarrow SFR)) \\ & + x(1-x)(1-e) \\ & (1-\exp[-R_0R(x,\infty)])^2(P(FFR \leftarrow IV) - P(IV \leftarrow FFR)),\end{aligned} \tag{4.21}$$

Efficiency model + IB-RA:

$$\begin{aligned}\frac{dx}{dt} = \; & x(1-x)\exp[-(1-\eta)R_0R(x,\infty)]\exp[-R_0R(x,\infty)] \\ & (P(SFR \leftarrow HV) - P(HV \leftarrow SFR)) \\ & + x(1-x)\exp[-(1-\eta)R_0R(x,\infty)] \\ & (1-\exp[-R_0R(x,\infty)])(P(FFR \leftarrow HV) - P(HV \leftarrow FFR)) \\ & + x(1-x)(1-\exp[-(1-\eta)R_0R(x,\infty)])\exp[-R_0R(x,\infty)] \\ & (P(SFR \leftarrow IV) - P(IV \leftarrow SFR)) \\ & + x(1-x)(1-\exp[-(1-\eta)R_0R(x,\infty)]) \\ & (1-\exp[-R_0R(x,\infty)])(P(FFR \leftarrow IV) - P(IV \leftarrow FFR)),\end{aligned} \tag{4.22}$$

Effectiveness model + SB-RA:

$$\begin{aligned}\frac{dx}{dt} = & -x(1-x)(e+(1-e)\exp[-R_0R(x,\infty)])P(HV \leftarrow NV) \\ & - x(1-x)(1-e)(1-\exp[-R_0R(x,\infty)])P(IV \leftarrow NV) \\ & + x(1-x)\exp[-R_0R(x,\infty)]P(SFR \leftarrow V) \\ & + x(1-x)(1-\exp[-R_0R(x,\infty)])P(FFR \leftarrow V),\end{aligned} \tag{4.23}$$

Efficiency model + SB-RA:

$$\begin{aligned}\frac{dx}{dt} = & -x(1-x)\exp[-(1-\eta)R_0R(x,\infty)]P(HV \leftarrow NV) \\ & - x(1-x)(1-\exp[-(1-\eta)R_0R(x,\infty)])P(IV \leftarrow NV) \\ & + x(1-x)\exp[-R_0R(x,\infty)]P(SFR \leftarrow V) \\ & + x(1-x)(1-\exp[-R_0R(x,\infty)])P(FFR \leftarrow NV),\end{aligned} \tag{4.24}$$

Effectiveness or efficiency model + DC:

$$\frac{dx}{dt} = -xP(V \leftarrow NV) + (1-x)P(NV \leftarrow V). \tag{4.25}$$

It is worthwhile to note that Eq. (4.25) is qualitatively consistent with what are called replicator dynamics (see; Eq. (2.6) in Sect. 2.1.1), one of the most common concepts in evolutionary game theory for expressing a system's dynamics. All of the above dynamical ODEs can be solved numerically. We introduce a so-called explicit scheme[32] for the time-varying terms to obtain a numerical solution; namely, vaccination coverage at equilibrium.

Figures 4.14 and 4.15 relating to the effectiveness and efficiency models, respectively, give the final epidemic size (left-hand panels), vaccination coverage (central panels), and average social payoff (right-hand panels) for various strategy-updating rules, namely IB-RA (upper panels), SB-RA (middle panels), and DC (lower panels).

The regions colored uniformly in light red (final epidemic size), dark blue (vaccination coverage), and light blue (average social payoff) indicate a pandemic taking place, where most individuals do not use vaccination (precisely speaking, not using either imperfect vaccination or an intermediate defense measure); thus, an almost-full-scale spread of infection occurs. Roughly speaking, these regions emerge when a smaller effectiveness (efficiency) is presumed or a larger cost is imposed. This seems quite natural because most individuals tend to shy away from vaccination if it is less reliable and/or too expensive. The border between each of these monotone regions and the remaining region implies a combination of critical effectiveness (efficiency) and critical vaccination cost to control the spread of an epidemic, causing an obvious change between the pandemic and controlled phases. As far as the controlled phase is concerned, interestingly, lower effectiveness (efficiency) can realize a higher vaccination coverage, which is also helped by the effect of lower cost. Even if a large fraction of individuals use vaccination, the epidemic cannot be eradicated because of the lower reliability of vaccination.

The detailed tendencies of the three updating rules differ, although the overall tendency is the same to some extent. Comparing the effectiveness model and the efficiency model, the latter has a wider pandemic phase at first glance. This implies that an intermediate defense measure with a certain η is less effective at suppressing the spread of an epidemic than imperfect vaccination, with e being defined as having the same numerical value as η.

To validate this theoretical framework, we conducted a series of numerical simulations based on the MAS approach discussed previously. Because we assumed a well-mixed population, we presumed a complete graph as the underlying network connecting the agents. Following previous studies, we set $\beta = 0.00088$, which was determined as the minimum transition rate β for which the final epidemic size exceeds the predefined threshold of 0.9 without vaccination. The results are shown in Figs. 4.16 and 4.17. Generally, all those results are respectively consistent with Figs. 4.14 and 4.15, although subtle discrepancies arise from the fact that the simulation presumed a finite population size of $N = 1000$.

[32]A reader can consult any standard textbooks on applied mathematics with numerical approaches. One example is provided below: Tanimoto (2014).

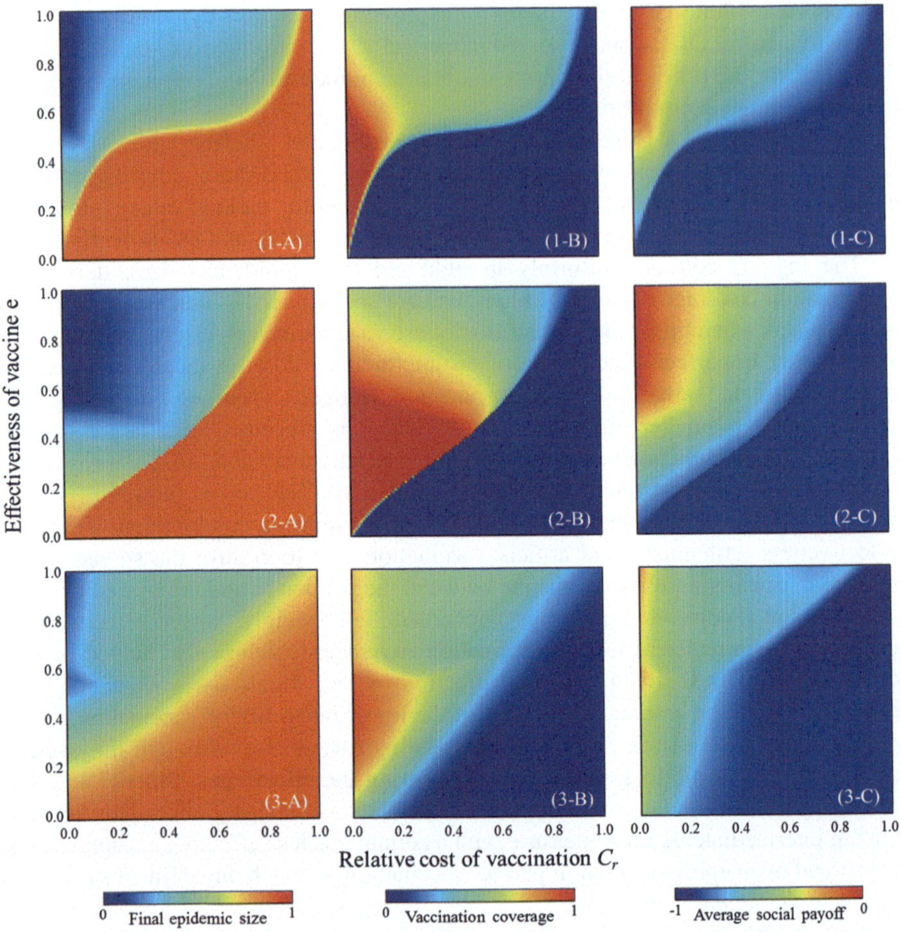

Fig. 4.14 Final epidemic size (left-hand panels; *-A), vaccination coverage (central panels; *-B), and average social payoff (right-hand panels; *-C) for strategy-updating rule IB-RA (upper panels; 1-*), SB-RA (middle panels; 2-*), and DC (lower panels; 3-*). The effectiveness model is applied

4.1.5.1 Summary

Presuming an infinite and well-mixed population, we established an analytical framework for a vaccination game in which three different strategy-updating rules were separately implemented. Our main concern was the extent to which the evolutionary picture differed when either imperfect vaccination or an intermediate defense measure was introduced to suppress the spread of an epidemic. We

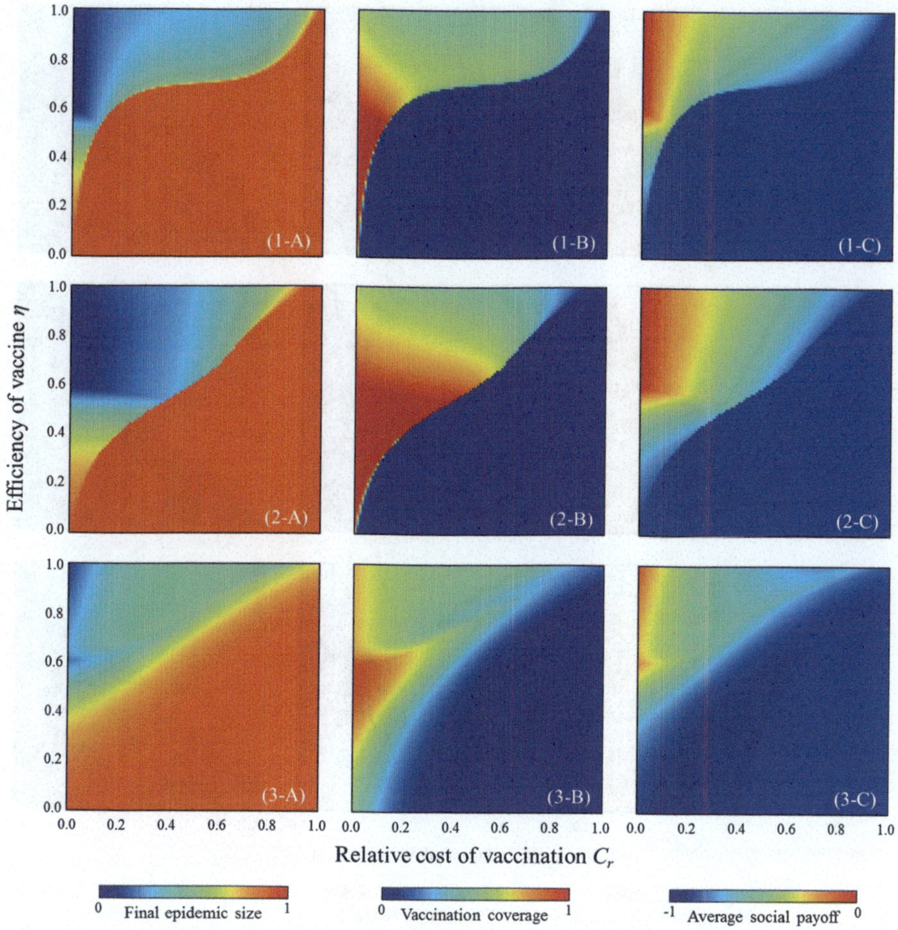

Fig. 4.15 Final epidemic size (left-hand panels; *-A), vaccination coverage (central panels; *-B), and average social payoff (right-hand panels; *-C) for strategy-updating rules IB-RA (upper panels; 1-*), SB-RA (middle panels; 2-*), and DC (bottom panels; 3-*). The efficiency model is applied

successfully established the respective evolutionary formulas, showing numerical results. We validated our framework by comparing its predictions with simulation results. As long as the same coefficient values for effectiveness and efficiency are presumed, an intermediate defense measure is marginally inferior to an imperfect vaccination.

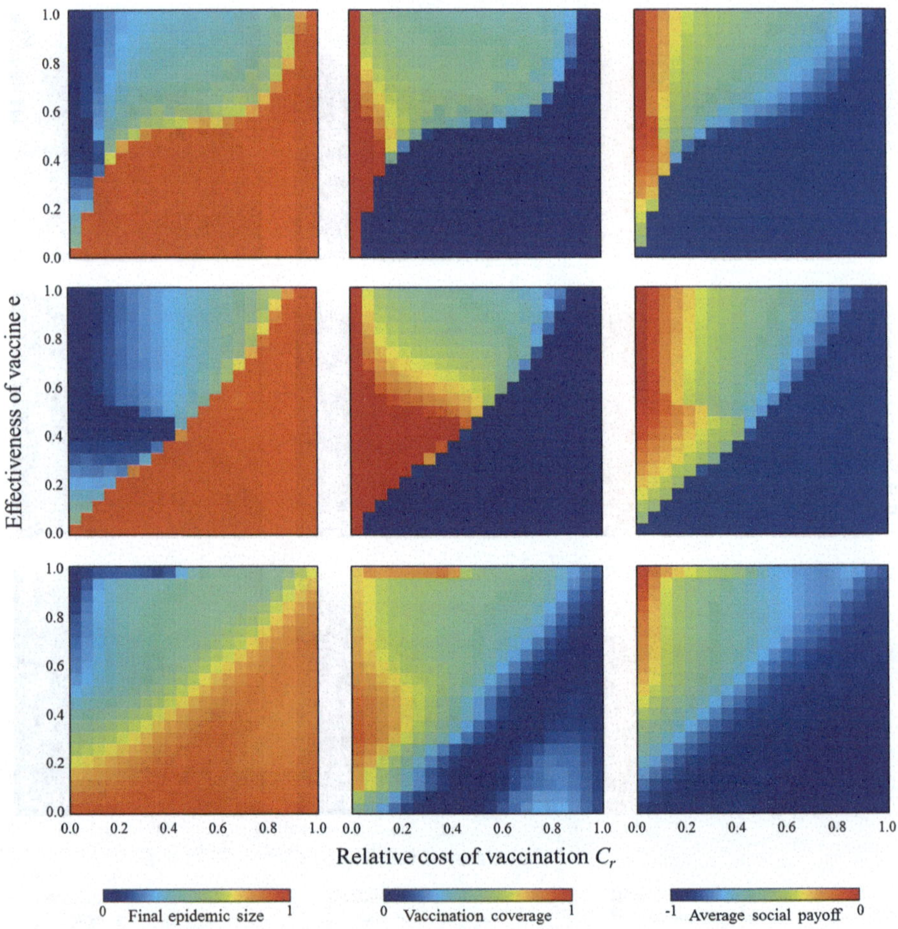

Fig. 4.16 MAS result for the final epidemic size (left-hand panels; *-A), vaccination coverage (central panels; *-B), and average social payoff (right-hand panels; *-C) for strategy-updating rules IB-RA (upper panels; 1-*), SB-RA (middle panels; 2-*), and DC (lower panels; 3-*). Effectiveness model is applied. We presumed a complete graph with $N = 1000$, $\beta = 0.00088$, and $\gamma = 1/3$

4.2 Optimal Subsidy-Policy Design for Vaccination

Many studies (called "vaccination games") have been reported on reproducing the vaccination dilemma. Meanwhile, returning to the social aspect of a voluntary vaccination policy, it may be possible to either solve or relax the vaccination dilemma. One provision is financial support from the central (or municipal) governments in the form of a subsidy. In fact, aged individuals and other people needing support are subsidized for the vaccination of seasonal influenza in Japan. On the other hand, the Japanese government is suffering from a huge budget deficit resulting from the stably increasing cost of medical care as well as the welfare cost

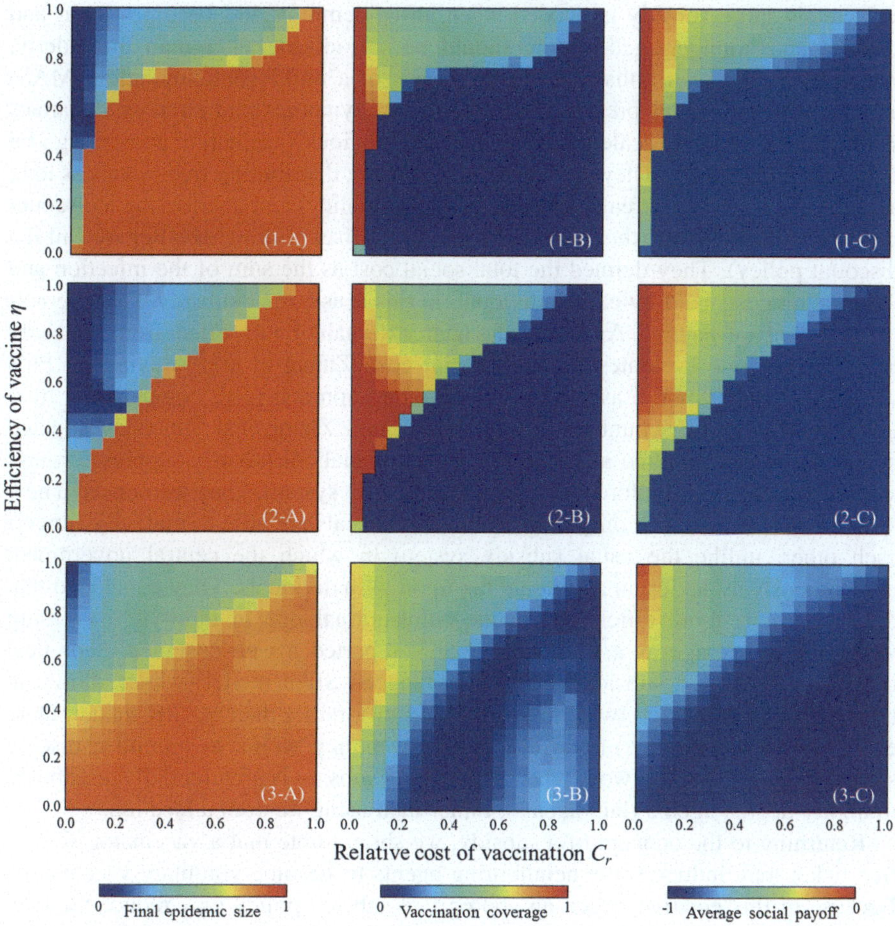

Fig. 4.17 MAS results for final epidemic size (left-hand panels; *-A), vaccination coverage (central panels; *-B), and average social payoff (right-hand panels; *-C) for strategy-updating rules IB-RA (upper panels; 1-*), SB-RA (middle panels; 2-*), and DC (lower panels; 3-*). The efficiency model is applied. We presumed a complete graph with $N = 1000$, $\beta = 0.00088$, and $\gamma = 1/3$

for an aged population. As long as a subsidy effectively reduces the entire social cost including the additional tax burden, such a subsidy policy for vaccination can be fully justified. Therefore, the question of what subsidy policy meeting this criterion can minimize the social cost is quite an important subject.

There have been several previous studies approaching this subject from different viewpoints. Gavious & Yamin[33] pioneered this field. Based on a macro model referring to SIR dynamics validated by field observations, they insisted that a

[33]Gavious and Yamin (2013).

sufficiently large subsidy is needed when an epidemic period becomes long, and middle- and younger-aged people should be subsidized rather than the elderly. Zhang et al.[34] gave a holistic report by means of a multiagent-simulation (MAS) approach. Their model presumed that epidemic dynamics and strategy dynamics share with a same time scale unlike most of the previous vaccination games (e.g., Fu et al. [10]) presuming. They compared two policies; distributing free-tickets as long as the budget allows (hereafter, called free-ticket policy) and distributing a discount ticket to all individuals (they called this a partial-offset policy; hereafter we call it a discount policy). They defined the total social cost as the sum of the infection and vaccination costs paid by each individual. In this sense, their subsidy was externalized as a "gift-of-nature". As we discuss later, externalizing the effect from a subsidy is justified if one evaluates the total social cost. Zhang et al.,[35] relying on Fu's vaccination-game model as well as an analytic approach [10], comparing a free-ticket policy with a discount policy for various cases. Zhang et al.[36] also investigated how random and targeted subsidization of individuals differently suppress disease spread. Tang et al.[37] explored an interesting subsidy system. They introduced a new specific system in which only neighboring individuals are able to mutually support each other, unlike the usual subsidy system in which the central government comprehensively levies an additional tax upon all individuals. They concluded that their system is more efficient than the conventional one in terms of increasing vaccination coverage as a whole. Li et al.[38] reported a well-designed theoretical model considering a subsidy system assuming an SIRS model (which allows an individual to be infected twice in a single season), rather than an SIR model. Ding et al.[39] highlighted how effectively a subsidy policy works to minimize disease spread on a scale-free network when a free-ticket subsidy is preferentially distributed to higher-degree agents (hub agents), rather than using random allocation.

Returning to the concept of a subsidy, we should note that a vaccinator given a free ticket may induce some neighboring agents to become voluntary vaccinators. Because of this positive effect, an appropriate subsidy policy may reduce the total social payoff as a whole. This seems analogous to the effect of a "stubborn vaccinator" investigated by Fukuda et al.[40] or a "zealot cooperator" in a spatial version of a 2 × 2 game, investigated Matsuzawa et al.,[41] meaning an agent insensitive to strategy updating. One important difference between a subsidized vaccinator and a stubborn vaccinator is whether or not their spatial location in a domain is frozen. Different people are given free tickets from one season to another.

---

[34]Zhang et al. (2013).

[35]Zhang et al. (2014).

[36]Zhang et al. (2017).

[37]Tang et al. (2017).

[38]Li et al. (2017).

[39]Ding et al. (2018).

[40]Fukuda and Tanimoto (2016).

[41]Matsuzawa et al. (2016).

Motivated by all of the aforementioned background, this section quantitatively answers the question of whether or not a vaccination-subsidy policy can really reduce the total social cost, comprising vaccination cost, diseases cost, and the tax burden needed to implement the subsidy. If this is possible, the question becomes what subsidy policy (e.g., distributing free or discount tickets) can minimize the total social cost. To do that, we, first off, establish a comprehensive vaccination-game model taking account of various subsidy mechanisms, and shed some light on what subsidy policy realizes the socially best solution. We use MAS as well as analytical approaches.

4.2.1 How We Model Subsidy Policy

Regardless of whether we take an MAS approach (see Sect. 4.1.3) or an analytic approach (See; Sect. 4.1.5), we follow the vaccination game depicted in the previous Section.

The key point is how we model various vaccination policies taken by the government.

From a policy-design point of view, we build four different subsidization procedures, which depend on whether a free-ticket or discount policy is presumed, and what target individuals are presumed.

Model A-1: Free-ticket policy. The target is as many randomly selected individuals as the subsidy budget will allow.

Model A-2: Free-ticket policy. The target is randomly selected individuals who did not vaccinate in the previous time step. As many tickets are distributed as the subsidy budget allows. Unlike Model A-1, this model targets potential non-vaccinators with the intent of increasing social-vaccination coverage.

Model A-3: Free-ticket policy. The target is randomly selected individuals who vaccinated in the previous time step. As many tickets are distributed as the subsidy budget allows. Unlike the previous two models, Model A-3 only gives the privilege of a free ticket to a cooperator (vaccinator) in the previous time step.

Model B: Discount policy. All individuals are given a discounting ticket to reduce the individual cost burden, C_r, by as much as possible. But this premier makes sense only when he decides committing vaccination.

Figure 4.18 illustrates the subsidy models above, where f_C is the cooperation rate (rate of vaccination) after each individual decides whether to vaccinate.

Let σ be the fraction of vaccinators to be subsidized out of the total population, N. Hence, the additional tax burden per capita for subsidies, TAX, is given as

$$TAX = \frac{C_r \cdot \sigma \cdot N}{N} = C_r \cdot \sigma. \tag{4.26}$$

Here, we define the total social payoff (TSP) as below:

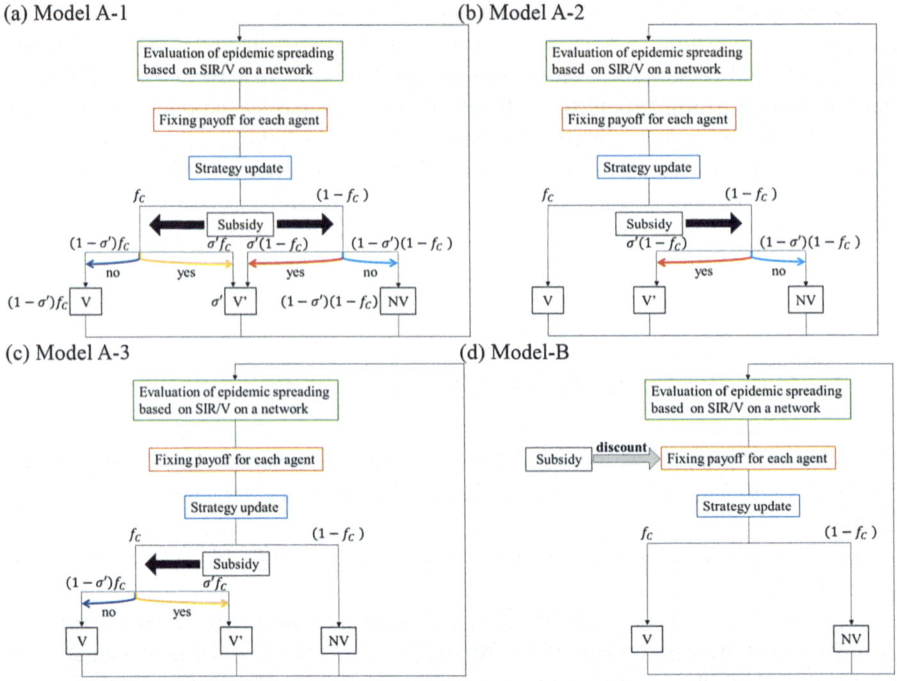

Fig. 4.18 Four models for subsidy policies; (**a**) Model A-1, (**b**) Model A-2, (**c**) Model A-3, and (**d**) Model-B

$$TSP = -C_r \cdot (VC - \sigma) - 1 \cdot FES - TAX = -C_r \cdot VC - 1 \cdot FES. \tag{4.27}$$

TSP must be negative, hence, |*TSP*| means the total social cost taking account of all aspects including the budget for a subsidy policy. It is worthwhile to note that *TSP* is independent of the size of a subsidy. This is quite natural, because governmental expenditure should be consistent with revenue as a whole. Eq. (4.27) revels that the subsidy works, in a sense, as a social "catalyst", allowing us to evaluate how much a subsidy (of a certain amount based on a certain policy) is able to enhance the number of voluntary vaccinators (or vaccinators in general). Thus, *TSP* can be thought of as an appropriate index for evaluating the social efficiency of a subsidy policy in the present study. We do not consider the additional cost burden levied on each individual, *TAX*, when evaluating each individual's payoff, because, in the real world, we cannot recognize exactly how much tax each individual has paid specifically for subsidies, as tax is generally levied. Thus, the breakdown of expenditure is not visible to us. However, a subsidy given as a free ticket or a discounting coupon is, of course, manifestly visible to the subsidized individual. This is why such a subsidy policy is generally favored and enthusiastically accepted by typical people in the street, on which the politics of "populism" tends to ride.

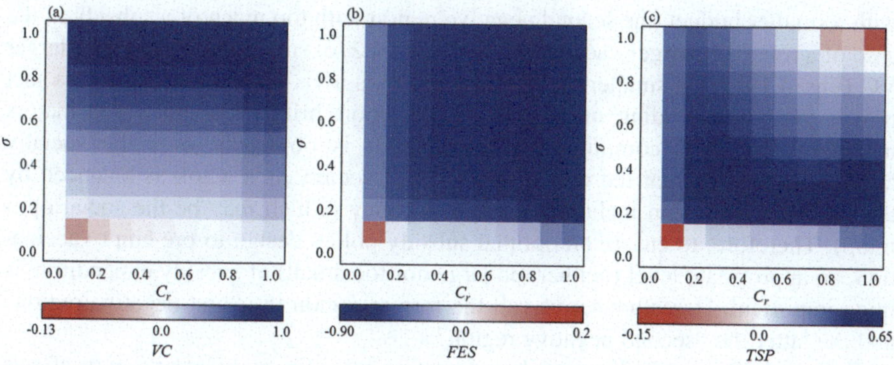

Fig. 4.19 Color indicates the difference between subsidy ($\sigma > 0$) and non-subsidy ($\sigma = 0$) cases. Panel (**a**) vaccination coverage (VC), (**b**) final epidemic size (FES), (**c**) total social payoff (TSP). Models A-1, lattice, and IB-RA are presumed

4.2.2 Results and Discussion; MAS Approach

We measure the vaccination coverage (VC), final epidemic size (FES), and total social payoff (TSP) with varying vaccination cost (C_r) and subsidy size (σ). To highlight how the introduction of a subsidy affects a social equilibrium, we consider the difference between cases with and without a subsidy. In the following visual results throughout this section, we are concerned on those differences for VC, FES and TSP. This, the three evaluation values respectively range from negative to positive. Hereafter, let us call a case without subsidy "default".

Figure 4.19 respectively shows the results for VC, FES, and TSP when presuming Model A-1 as a subsidy setting, where a free ticket is distributed to randomly selected individuals regardless of V or NV, using a lattice for an underlying network and IB-RA for a strategy-updating rule. Red indicates a region where a subsidy devastates social efficiency, leading to a smaller fraction of vaccinators, larger final epidemic size, and smaller total social payoff than in the default case.

Let us consider the total social payoff (Fig. 4.19c), where there are two negative regions. One occurs when a smaller vaccination cost and subsidy budget are presumed. Another takes place when a larger vaccination cost and subsidy budget are presumed. The former accords with regions of negative-VC difference (Fig. 4.19a) and positive-FES difference (Fig. 4.19b); in this region, subsidies going to defective individuals who have no intention of vaccinating eventually reduce the number of voluntary vaccinators (e.g., self-financed vaccinators), leading to a wider spread of disease and pushing up the total social cost due to a larger number of infectious individuals. This is quite ironic from the viewpoint of social context. The implication drawn from this particular case is that too small of a budget for the subsidy policy in the case of a relatively smaller vaccination cost fails to increase the number of voluntary vaccinators, but rather causes individuals to refrain from cooperating in establishing a herd immunity. Unlike the first negative region with a smaller budget, the second negative region with too much of a subsidy in the case of a relatively larger vaccination cost (Fig. 4.27c) successfully realizes a larger VC (Fig. 4.19a) and smaller FES (Fig. 4.19b) vis-à-vis the default case. This fact proves that overspending on the financial support brings too much vaccinators including forcefully committing individuals. It is obviously over the socially approved level whereat the cost of infection of society as a whole is balanced by the cost of vaccination including a subsidy policy (which may be the social optimum). Therefore, to ensure an optimal subsidy-policy design to pre-empt vaccination, an appropriate level (neither too large nor too small) of subsidy expenditure is quite important. Hereafter, let us call the former region the "first negative region" and the latter the "second negative region".

Figures 4.20 and 4.21 illustrate how targeting subsidies at different groups affects the total social cost in comparison to Fig. 4.19. Interestingly, Model A-3 (Fig. 4.21), where financial support is given only to cooperative individuals, does not incur the first negative region of TSP difference. This suggests that, to increase the number of voluntary vaccinators, helping potential vaccinators rather than helping potential social-defectors aiming to free-ride on herd immunity would be more beneficial. Relating to this, comparison of Fig. 4.19 with Fig. 4.20 implies that the result of helping potential social defectors brings almost no difference from that of targeting any arbitrary people.

Figure 4.22 shows the result when the Model-B discount policy is presumed. This should be compared with Fig. 4.21 on the grounds that target individuals are potential cooperators (the discount becomes meaningful only when an individual voluntarily vaccinates). The result shows no specific difference between those two policies. As long as the intent is to increase the number of voluntary vaccinators in a society, a "priority system" distributing free tickets to some potential vaccinators does not differ in effect from an "egalitarian system" allowing a certain discount (distributing discount coupons) to all potential vaccinators.

Figure 4.23 highlights the difference in strategy updating. This should be compared with Fig. 4.19. With presuming SB-RA relying on global information when updating strategy, the first negative region with a small budget is slightly extended compared with IB-RA relying on local information (only referring to the neighbors' payoff), whereas the second negative region with a too-high budget shrinks slightly vis-à-vis IB-RA. Yet, as a whole, the difference between those two strategy-updating rules looks small, which indicates whether an individual decision referring to global or only local information makes a less-significant difference in the social cost.

Figures 4.24 and 4.25 compare the topological contributions. Figure 4.24 presumes RRG while Fig. 4.25 show the case of BA-SF, which should be also compared with Fig. 4.19. The comparison between Fig. 4.19 and Fig. 4.24 confirms that the effect of random links results in a quite large difference from the case of a regular & homogeneous network (lattice; Fig. 4.19). The existence of random links devastates the efficiency of subsidy policy, resulting from the significantly wider negative regions compared with those observed in Fig. 4.19. In particular, the first negative region with a smaller budget spreads to much larger C_r as well as larger σ. Moreover, comparing RRG (Fig. 4.24) with BA-SF (Fig. 4.25), we should note that

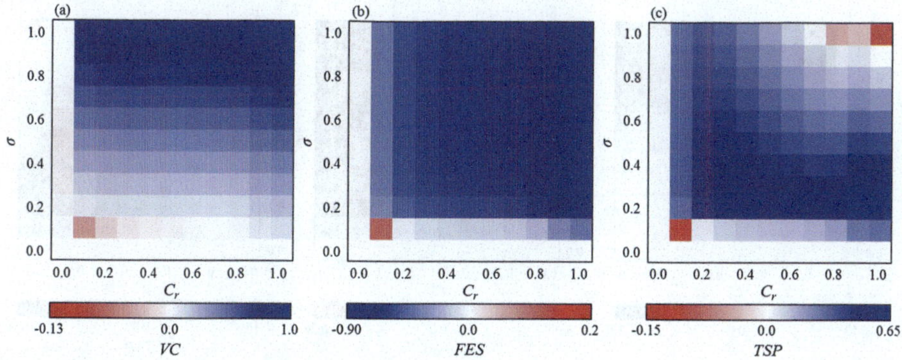

Fig. 4.20 Color indicates the difference between subsidy ($\sigma > 0$) and non-subsidy ($\sigma = 0$) cases. Panel (**a**) vaccination coverage (VC), (**b**) final epidemic size (FES), (**c**) total social payoff (TSP). Models A-2, lattice, and IB-RA are presumed

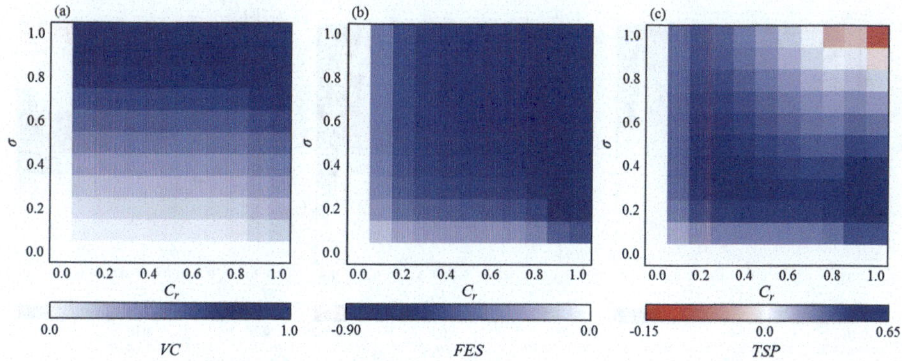

Fig. 4.21 Color indicates the difference between subsidy ($\sigma > 0$) and non-subsidy ($\sigma = 0$) cases. Panel (**a**) vaccination coverage (VC), (**b**) final epidemic size (FES), (**c**) total social payoff (TSP). Models A-3, lattice, and IB-RA are presumed

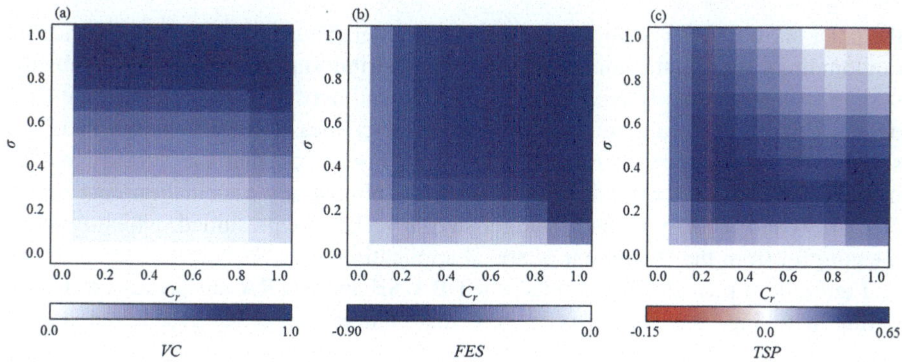

Fig. 4.22 Color indicates the difference between subsidy ($\sigma > 0$) and non-subsidy ($\sigma = 0$) cases. Panel (**a**) vaccination coverage (VC), (**b**) final epidemic size (FES), (**c**) total social payoff (TSP). Models B, lattice, and IB-RA are presumed

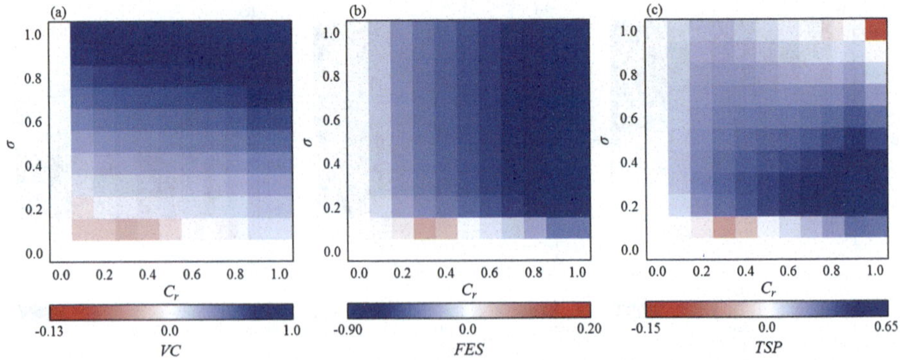

Fig. 4.23 Color indicates the difference between subsidy ($\sigma > 0$) and non-subsidy ($\sigma = 0$) cases. Panel (**a**) vaccination coverage (VC), (**b**) final epidemic size (FES), (**c**) total social payoff (TSP). Models A-1, lattice, and SB-RA are presumed

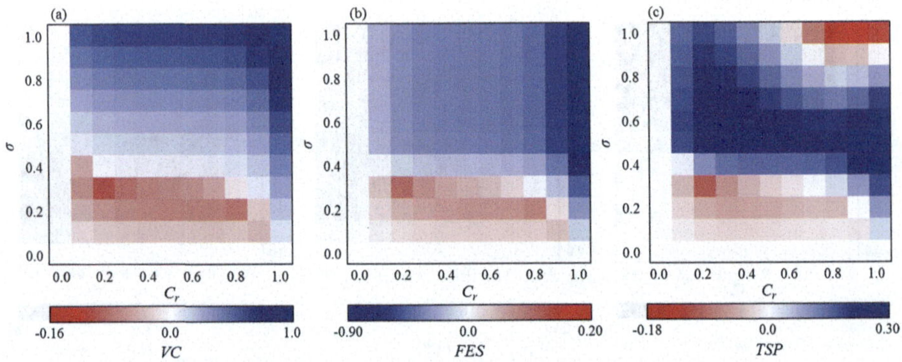

Fig. 4.24 Color indicates the difference between subsidy ($\sigma > 0$) and non-subsidy ($\sigma = 0$) cases. Panel (**a**) vaccination coverage (VC), (**b**) final epidemic size (FES), (**c**) total social payoff (TSP). Models A-1, RRG, and IB-RA are presumed

the latter shows worse performance than the former. Thus, not only the existence of random links but also the non-uniform degree distribution undermines to the subsidy system. As is commonly recognized, human social networks can feature scale-free degree distributions, unlike small-world characteristics.[42] Recalling this fact, we should note the important result observed in Fig. 4.25c that the subsidy works inversely to reduce the total social cost when almost any vaccination-cost range ($C_r < 0.9$) and feasibly realistic budget size ($\sigma < 0.5$) are presumed, which may be meaningful from the standpoint of social applications.

Figure 4.26 provides the results when BA-SF and SB-RA are presumed. Comparing with Fig. 4.25, we should note that, unlike the lattice case (comparison

[42]Masuda (2017).

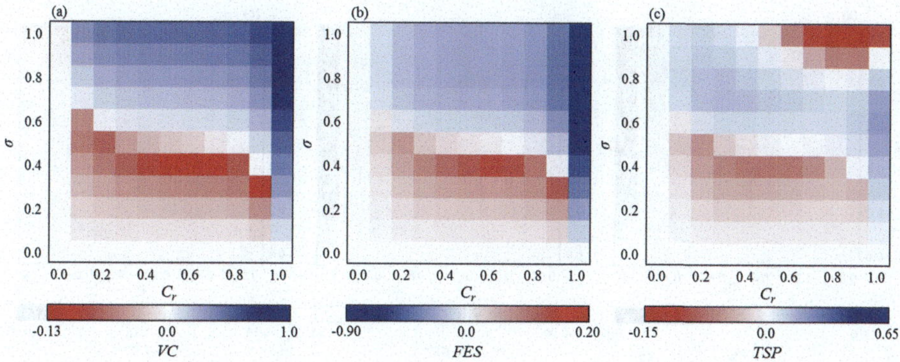

Fig. 4.25 Color indicates the difference between subsidy ($\sigma > 0$) and non-subsidy ($\sigma = 0$) cases. Panel (**a**) vaccination coverage (VC), (**b**) final epidemic size (FES), (**c**) total social payoff (TSP). Models A-1, BA-SF, and IB-RA are presumed

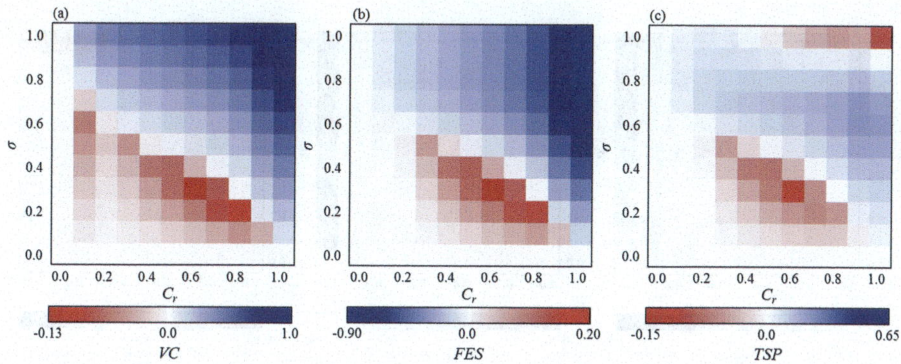

Fig. 4.26 Color indicates the difference between subsidy ($\sigma > 0$) and non-subsidy ($\sigma = 0$) cases. Panel (**a**) vaccination coverage (VC), (**b**) final epidemic size (FES), (**c**) total social payoff (TSP). Models A-1, BA-SF, and SB-RA are presumed

between Figs. 4.19 and 4.23), BA-SF causes a relatively significant difference between the two different update rules. Contrasting with the lattice case, the introduction of global information (Fig. 4.26) somewhat relaxes the drawbacks of subsidy introduction. In particular, the first negative region with a relatively smaller budget is shrunken compared with the counterpart region in Fig. 4.25.

We show the free-ticket and discount-coupon cases with BA-SF being presumed as an underlying network, as shown in Figs. 4.27 and 4.28 (where IB-RA is imposed). Between those two, likewise the lattice case does (comparison between Figs. 4.21 and 4.22), there is no phenomenal difference. But more importantly, as the lattice showing as well, the first negative region, where a relatively smaller budget size rather devastating the social efficiency than the default case, does disappear. Although, as mentioned above, a subsidy policy presuming a BA-SF network may work badly in terms of social efficiency vis-a-vis the lattice case, it would be significantly improved

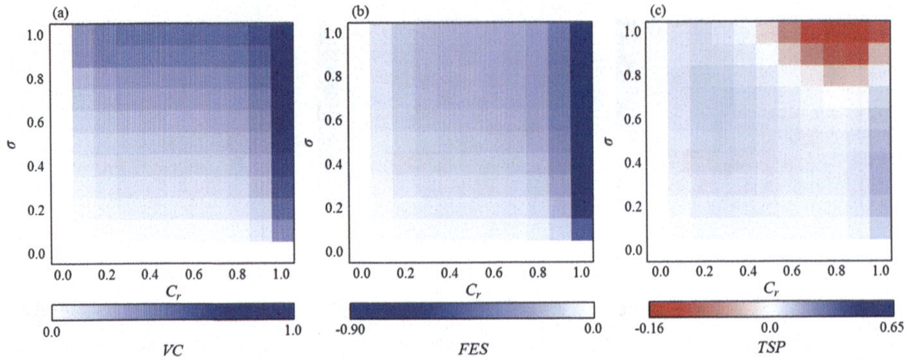

Fig. 4.27 Color indicates the difference between subsidy ($\sigma > 0$) and non-subsidy ($\sigma = 0$) cases. Panel (**a**) vaccination coverage (VC), (**b**) final epidemic size (FES), (**c**) total social payoff (TSP). Models A-3, BA-SF, and IB-RA are presumed

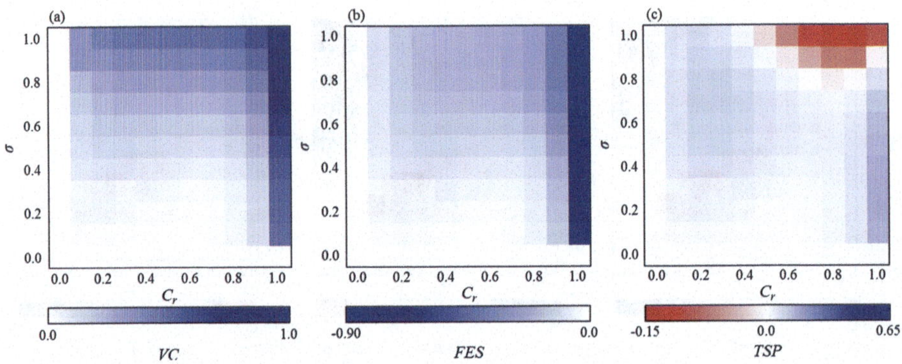

Fig. 4.28 Color indicates the difference between subsidy ($\sigma > 0$) and non-subsidy ($\sigma = 0$) cases. Panel (**a**) vaccination coverage (VC), (**b**) final epidemic size (FES), (**c**) total social payoff (TSP). Models B, BA-SF, and IB-RA are presumed

if the government were to rely on the principle of "heaven helps those who help themselves"; e.g., a policy distributing free tickets only to potential vaccinators or distributing discount coupons that are only valid to those who actually vaccinate. This seems quite important from a social-application point of view.

4.2.3 Results and Discussion: Analytic Approach

We utilize the modified SIR model, which can reproduce the epidemic dynamics arising from imperfect vaccination, namely the effectiveness model introduced by our previous work Ref [13]. In the present study, we further modify for the vaccination game under a subsidy policy. We assume a population that is infinite and ideally well mixed, implying that the total population of N in the MAS model is normalized as $N = 1$. This suggests that the individual tax burden for a subsidy denoted by TAX in Eq. (4.26) is equivalent to the exact size of the subsidy budget (hereafter denoted by SB) in the analytical model. Namely, to connect the MAS model with the analytical model, we should note

$$TAX = \frac{C_r \cdot \sigma \cdot N}{N} = C_r \cdot \sigma = C_r \cdot \sigma \cdot 1 = SB. \tag{4.28}$$

4.2.3.1 Epidemic Dynamics

A vaccinated population is separated into two classes: immune individuals obtaining perfect immunity and non-immune ones failing to obtain immunity. Let the effectiveness of the vaccine and the vaccination coverage be e ($0 \leq e \leq 1$) and x, respectively. The effectiveness refers to a situation in which some vaccinated agents acquire immunity with effectiveness probability e; meanwhile, the remaining agents fail to acquire immunity with probability $1 - e$. The fraction of vaccinated individuals with immunity must be ex, while that of non-immune individuals is $(1 - ex)$. We can express the final epidemic size, R, in relation to both x and time t at equilibrium ($t = \infty$) as

$$R(x, \infty) = (1 - ex)(1 - \exp[-R_0 R(x, \infty)]), \tag{4.29}$$

where R_0 is the basic reproduction number. $R(x, \infty)$ gives the respective fractions of four different types of individual depending on whether they are vaccinated or non-vaccinated and healthy or infected, as summarized in Table 4.4.

4.2.3.2 Payoff Structure

Again, an epidemic season continues until all infected individuals recover, meaning that the number of infected individuals is zero. As discussed in Sect. 4.1., we put the individual cost burden to support a subsidy system aside. If non-vaccinated individuals are infected, they incur an infection cost of 1. By contrast, non-vaccinated individuals who fortunately remain healthy can avoid the cost burden. Moreover, individuals who unfortunately are infected despite committing vaccination must pay the cost $C_r + 1$. Consequently, the payoff of each individual at the end of an epidemic season depends upon their final state. Table 4.5 summarizes the payoff whether committing to a provision (either vaccination or defense against contagion) or not and whether remaining healthy or becoming infected.

Table 4.4 Fractions of four types of individual using the effectiveness model without subsidy policy

Strategy/state	Healthy	Infected
Vaccinated	$x(e + (1 - e)\exp[-R_0 R(x, \infty)])$	$x(1 - e)(1 - \exp[-R_0 R(x, \infty)])$
Non-vaccinated	$(1 - x)\exp[-R_0 R(x, \infty)]$	$(1 - x)(1 - \exp[-R_0 R(x, \infty)])$

Table 4.5 Payoff structure determined at the end of an epidemic season

Strategy/state	Healthy	Infected
Vaccinated	$-C_r$	$-C_r - 1$
Non-vaccinated	0	-1

4.2.3.3 Subsidy Policies

As depicted in Sect. 4.2.1, we presume four different subsidy policies; Models A-1, A-2, A-3, and B. In the former three models, the limited subsidy is randomly distributed after each individual decides whether to vaccinate (see Fig. 4.18). In Model A-2, the defectors who decided not to vaccinate randomly receive the subsidy and vaccinate without vaccination cost. By contrast, in Model A-3, the cooperators who decide to vaccinate with their own expense can cover the vaccination cost by a subsidy. In Model A-1, the subsidy is randomly distributed to the population irrespective of whether individuals are defectors or cooperators. In these frameworks, a subsidized individual definitely vaccinates without a vaccination cost irrespective of his decision because he is given a free ticket. Hence, the subsidy refunds the voluntary vaccination cost for a cooperator, and a free ticket literally works for a defector. In the discount policy, Model B, all vaccinators can receive a certain amount of a subsidy, which reduces the vaccination cost. One point to note is that the acting discount rate by the subsidy depends on the vaccination coverage at a certain time-step.

Model A-1

In this scenario, the total subsidy is distributed to a certain fraction of individuals. Let this fraction be σ'. In Model A-1, $\sigma' = SB/C_r$ indicates that σ' happens to be consistent with σ if one refers to Eq. (4.27); $\sigma = SB/C_r$. Let the cooperation (defection) rate be f_C ($f_D = 1 - f_C$). All subsidized cooperators $\sigma' f_C$ and all subsidized defectors $\sigma'(1 - f_C)$ take vaccination without personal cost. Therefore, the vaccination coverage is $x = f_C + \sigma'(1 - f_C)$. We can find the respective fractions of eight different types of individual depending on whether they are vaccinated, non-vaccinated, or subsidized and whether they are healthy or infected, as summarized in Table 4.6. We also can present the payoff structure in Table 4.7 instead of Table 4.5.

Table 4.6 Fractions of eight types of individual using effectiveness model A-1

		Healthy	Infected
C	Vaccinated	$(1-\sigma')f_C(e+(1-e)\exp[-R_0R(x,\infty)])$	$(1-\sigma')f_C(1-e)(1-\exp[-R_0R(x,\infty)])$
	Subsidized	$\sigma' f_C(e+(1-e)\exp[-R_0R(x,\infty)])$	$\sigma' f_C(1-e)(1-\exp[-R_0R(x,\infty)])$
D	Non-vaccinated	$(1-\sigma')(1-f_C)\exp[-R_0R(x,\infty)]$	$(1-\sigma')(1-f_C)(1-\exp[-R_0R(x,\infty)])$
	Subsidized	$\sigma'(1-f_C)(e+(1-e)\exp[-R_0R(x,\infty)])$	$\sigma'(1-f_C)(1-e)(1-\exp[-R_0R(x,\infty)])$

Table 4.7 Payoff structure presuming Model A-1

		Healthy	Infected
C	Vaccinated	$-C_r$	$-C_r-1$
	Subsidized	0	-1
D	Non-vaccinated	0	-1
	Subsidized	0	-1

Subsequently, we can evaluate the expected payoffs in the form of the average social payoff $\langle\pi\rangle$, the average corporative payoff $\langle\pi_C\rangle$, and the average defective payoff $\langle\pi_D\rangle$ for the imperfect vaccination:

$$\begin{aligned}\langle\pi\rangle = \; & -C_r(1-\sigma')f_C(e+(1-e)\exp[-R_0R(x,\infty)]) \\ & -(C_r+1)(1-\sigma')f_C(1-e)(1-\exp[-R_0R(x,\infty)]) \\ & -\sigma' f_C(1-e)(1-\exp[-R_0R(x,\infty)]) - (1-\sigma')(1-f_C)(1-\exp[-R_0R(x,\infty)]) \\ & -\sigma'(1-f_C)(1-e)(1-\exp[-R_0R(x,\infty)]) - SB,\end{aligned} \tag{4.30}$$

$$\begin{aligned}\langle\pi_C\rangle = \; & -C_r(1-\sigma')(e+(1-e)\exp[-R_0R(x,\infty)]) \\ & -(C_r+1)(1-\sigma')(1-e)(1-\exp[-R_0R(x,\infty)]) \\ & -\sigma'(1-e)(1-\exp[-R_0R(x,\infty)]),\end{aligned} \tag{4.31}$$

$$\begin{aligned}\langle\pi_D\rangle = & -(1-\sigma')(1-\exp[-R_0R(x,\infty)]) \\ & -\sigma'(1-e)(1-\exp[-R_0R(x,\infty)]).\end{aligned} \tag{4.32}$$

It is worth noting that $\langle\pi\rangle$ is not consistent with the fraction weighted sum of $\langle\pi_C\rangle$ and $\langle\pi_D\rangle$, because $\langle\pi\rangle$ accounts –*SB*. As mentioned in Sect. 4.2.1, the total social payoff, *TSP*, which is consistent with $\langle\pi\rangle$ must consider the tax burden to support the subsidy. Unlike Eq. (4.27), Eq. (4.30) explicitly contains the tax term –*SB*, which is attributed to the fact that we did not account for the cost burden at an individual level, as we assumed in Sect. 4.2.1.

Model A-2

In this scenario, the total amount of a subsidy is distributed to a certain fraction of defectors. Let this fraction be σ'. In Model A-2, note that $\sigma' = SB/C_r(1-f_C)$. All subsidized defectors $\sigma'(1-f_C)$ take vaccination without personal cost. Therefore, the vaccination coverage is $x = f_C + \sigma'(1-f_C)$. We can obtain the respective fractions of

Table 4.8 Fractions of six types of individual using the effectiveness model A-2

		Healthy	Infected
C	Vaccinated	$f_C(e + (1 - e) \exp[-R_0R(x,\infty)])$	$f_C(1 - e)(1 - \exp[-R_0R(x,\infty)])$
D	Non-vaccinated	$(1 - \sigma')(1 - f_C) \exp[-R_0R(x,\infty)]$	$(1 - \sigma')(1 - f_C)(1 - \exp[-R_0R(x,\infty)])$
	Subsidized	$\sigma'(1 - f_C)(e + (1 - e) \exp[-R_0R(x,\infty)])$	$\sigma'(1 - f_C)(1 - e)(1 - \exp[-R_0R(x,\infty)])$

Table 4.9 Payoff structure with presuming Model A-2

		Healthy	Infected
C	Vaccinated	$-C_r$	$-C_r - 1$
D	Non-vaccinated	0	-1
	Subsidized	0	-1

Table 4.10 Fractions of six types of individual using effectiveness model A-3

		Healthy	Infected
C	Vaccinated	$(1 - \sigma')f_C(e + (1 - e) \exp[-R_0R(x,\infty)])$	$(1 - \sigma')f_C(1 - e)(1 - \exp[-R_0R(x,\infty)])$
	Subsidized	$\sigma' f_C(e + (1 - e) \exp[-R_0R(x,\infty)])$	$\sigma' f_C(1 - e)(1 - \exp[-R_0R(x,\infty)])$
D	Non-vaccinated	$(1 - f_C) \exp[-R_0R(x,\infty)]$	$(1 - f_C)(1 - \exp[-R_0R(x,\infty)])$

six different types of individual depending on whether they are vaccinated, non-vaccinated, or subsidized and whether they are healthy or infected, as summarized in Table 4.8. We also can present the payoff structure in Table 4.9.

We can evaluate the expected payoffs in the form of the average social payoff $\langle \pi \rangle$, the average cooperative payoff $\langle \pi_C \rangle$, and the average defective payoff $\langle \pi_D \rangle$ for imperfect vaccination as

$$\begin{aligned}\langle \pi \rangle = \; & -C_r f_C(e + (1-e)\exp[-R_0R(x,\infty)]) \\ & -(C_r+1)f_C(1-e)(1-\exp[-R_0R(x,\infty)]) \\ & -(1-\sigma')(1-f_C)(1-\exp[-R_0R(x,\infty)]) \\ & -\sigma'(1-f_C)(1-e)(1-\exp[-R_0R(x,\infty)]) \\ & -SB,\end{aligned} \tag{4.33}$$

$$\begin{aligned}\langle \pi_C \rangle = & -C_r(e + (1-e)\exp[-R_0R(x,\infty)]) \\ & - (C_r+1)(1-e)(1-\exp[-R_0R(x,\infty)]),\end{aligned} \tag{4.34}$$

$$\begin{aligned}\langle \pi_D \rangle = & -(1-\sigma')(1-\exp[-R_0R(x,\infty)]) \\ & - \sigma'(1-e)(1-\exp[-R_0R(x,\infty)]).\end{aligned} \tag{4.35}$$

Model A-3

In this scenario, the total subsidy amount is distributed to a certain fraction of cooperators. Let this fraction be σ'. In Model A-3, note $\sigma' = SB/C_r f_C$. All subsidized

Table 4.11 Payoff structure presuming Model A-3

		Healthy	Infected
C	Vaccinated	$-C_r$	$-C_r - 1$
	Subsidized	0	-1
D	Non-vaccinated	0	-1

cooperators $\sigma' f_C$ take vaccination without personal cost. We can obtain the fractions of six different types of individual depending on whether they are vaccinated, non-vaccinated, or subsidized and whether they are healthy or infected, as summarized in Table 4.10. We also can present the payoff structure in Table 4.11.

We can evaluate the expected payoffs in the form of the average social payoff $\langle \pi \rangle$, the average cooperative payoff $\langle \pi_C \rangle$, and the average defective payoff $\langle \pi_D \rangle$ for imperfect vaccination as

$$\begin{aligned}\langle \pi \rangle = \; & -C_r(1-\sigma')f_C(e+(1-e)\exp[-R_0R(x,\infty)]) \\ & -(C_r+1)(1-\sigma')f_C(1-e)(1-\exp[-R_0R(x,\infty)]) \\ & -\sigma' f_C(1-e)(1-\exp[-R_0R(x,\infty)]) \\ & -(1-f_C)(1-\exp[-R_0R(x,\infty)])-SB,\end{aligned} \tag{4.36}$$

$$\begin{aligned}\langle \pi_C \rangle = \; & -C_r(1-\sigma')(e+(1-e)\exp[-R_0R(x,\infty)]) \\ & -(C_r+1)(1-\sigma')(1-e)(1-\exp[-R_0R(x,\infty)]) \\ & -\sigma'(1-e)(1-\exp[-R_0R(x,\infty)]),\end{aligned} \tag{4.37}$$

$$\langle \pi_D \rangle = -(1-\exp[-R_0R(x,\infty)]). \tag{4.38}$$

Model B

Under the discount-subsidy policy, vaccination coverage is equal to the cooperation rate because the subsidy is distributed to all vaccinated individuals. Hence, the fractions of four types of individual are the same as in Table 4.4. The total amount of subsidy is equally distributed to all vaccinators, and a vaccinator can reduce the vaccination cost by SB/f_C. Thus, we obtain Table 4.12 for the modified payoff structure.

We can evaluate the expected payoffs in the form of the average social payoff $\langle \pi \rangle$, the average cooperative payoff $\langle \pi_C \rangle$, and the average defective payoff $\langle \pi_D \rangle$ for imperfect vaccination as

$$\begin{aligned}\langle \pi \rangle = \; & \left(-C_r+\frac{SB}{f_C}\right)f_C(e+(1-e)\exp[-R_0R(x,\infty)]) \\ & +\left(-C_r-1+\frac{SB}{f_C}\right)f_C(1-e)(1-\exp[-R_0R(x,\infty)]) \\ & -(1-f_CS)(1-\exp[-R_0R(x,\infty)])-SB,\end{aligned} \tag{4.39}$$

Table 4.12 Payoff structure under the cooperator-preferential subsidy policy

		Healthy	Infected
C	Subsidized	$-C_r + \frac{SB}{f_C}$	$-C_r - 1 + \frac{SB}{f_C}$
D	Non-vaccinated	0	-1

$$\langle \pi_C \rangle = \left(-C_r + \frac{SB}{f_C} \right)(e + (1-e)\exp[-R_0 R(x,\infty)]) + \left(-C_r - 1 + \frac{SB}{f_C} \right)(1-e)(1 - \exp[-R_0 R(x,\infty)]), \tag{4.40}$$

$$\langle \pi_D \rangle = -(1 - \exp[-R_0 R(x,\infty)]). \tag{4.41}$$

4.2.3.4 Strategy Adaptation

Following Sect. 4.1.4, we specify how IB-RA and SB-RA are embedded in our theoretical framework.

IB-RA

We refer to Eq. (4.14). In the present framework, there are eight possible classes of individual state in relation to the cost burden: (i) a healthy defector (HD), who pays nothing, (ii) an infected defector (ID), who pays -1, (iii) an infected cooperator (IC), who pays $-C_r - 1$, (iv) a healthy cooperator (HC), who pays $-C_r$, (v) and (vi) a healthy subsidized defector and cooperator (HSD, HSC), respectively, who pay nothing due to a subsidy covering vaccination cost, and (vii) and (viii) an infected subsidized defector and cooperator (ISD, ISC, respectively) who pay -1 due to infection but pay nothing due to subsidy covering the vaccination cost. Each individual has two strategies: vaccination (V) (i.e., cooperation, C) and non-vaccination (NV) (i.e., defection, D). Thus, the transition probability that affects the time transition of f_C, which should be considered in the IB-RA rule, is covered by one of the following 32 cases:

$$P(HC \leftarrow HD) = P(HC \leftarrow HSD) = \frac{1}{1 + \exp[-(0 - (-C_r))/\kappa]}, \tag{4.42a}$$

$$P(HC \leftarrow ID) = P(HC \leftarrow ISD) = \frac{1}{1 + \exp[-(-1 - (-C_r))/\kappa]}, \tag{4.42b}$$

$$P(IC \leftarrow HD) = P(IC \leftarrow HSD) = \frac{1}{1+\exp[-(0-(-C_r-1))/\kappa]}, \tag{4.42c}$$

$$P(IC \leftarrow ID) = P(IC \leftarrow ISD) = \frac{1}{1+\exp[-(-1-(-C_r-1))/\kappa]}, \tag{4.42d}$$

$$P(HCS \leftarrow HD) = P(HCS \leftarrow HSD) = \frac{1}{1+\exp[-(0-0)/\kappa]}, \tag{4.42e}$$

$$P(HCS \leftarrow ID) = P(HCS \leftarrow ISD) = \frac{1}{1+\exp[-(-1-0)/\kappa]}, \tag{4.42f}$$

$$P(ICS \leftarrow HD) = P(ICS \leftarrow HSD) = \frac{1}{1+\exp[-(0-(-1))/\kappa]}, \tag{4.42g}$$

$$P(ICS \leftarrow ID) = P(ICS \leftarrow ISD) = \frac{1}{1+\exp[-(-1-(-1))/\kappa]}, \tag{4.42h}$$

$$P(HD \leftarrow HC) = P(HSD \leftarrow HC) = \frac{1}{1+\exp[-(-C_r-0)/\kappa]}, \tag{4.42i}$$

$$P(HD \leftarrow IC) = P(HSD \leftarrow IC) = \frac{1}{1+\exp[-((-C_r-1)-0)/\kappa]}, \tag{4.42j}$$

$$P(HD \leftarrow HCS) = P(HSD \leftarrow HCS) = \frac{1}{1+\exp[-(0-0)/\kappa]}, \tag{4.42k}$$

$$P(HD \leftarrow ICS) = P(HSD \leftarrow ICS) = \frac{1}{1+\exp[-((-1)-0)/\kappa]}, \tag{4.42l}$$

$$P(ID \leftarrow HC) = P(ISD \leftarrow HC) = \frac{1}{1+\exp[-(-C_r-(-1))/\kappa]}, \tag{4.42m}$$

$$P(ID \leftarrow IC) = P(ISD \leftarrow IC) = \frac{1}{1+\exp[-((-C_r-1)-(-1))/\kappa]}, \tag{4.42n}$$

$$P(ID \leftarrow HCS) = P(ISD \leftarrow HCS) = \frac{1}{1+\exp[-(0-(-1))/\kappa]}, \tag{4.42o}$$

$$P(ID \leftarrow ICS) = P(ISD \leftarrow ICS) = \frac{1}{1+\exp[-((-1)-(-1))/\kappa]}. \tag{4.42p}$$

When Model B is presumed, only the following probabilities are possible;

$$P(HSC \leftarrow HD) = \frac{1}{1+\exp[-(0-(-C_r+SB/f_C))/\kappa]}, \tag{4.43a}$$

$$P(HSC \leftarrow ID) = \frac{1}{1 + \exp[-(-1 - (-C_r + SB/f_C))/\kappa]}, \tag{4.43b}$$

$$P(ISC \leftarrow HD) = \frac{1}{1 + \exp[-(0 - (-C_r - 1 + SB/f_C))/\kappa]}, \tag{4.43c}$$

$$P(ISC \leftarrow ID) = \frac{1}{1 + \exp[-(-1 - (-C_r - 1 + SB/f_C))/\kappa]}, \tag{4.43d}$$

$$P(HD \leftarrow HSC) = \frac{1}{1 + \exp[-(-C_r + SB/f_C - 0)/\kappa]}, \tag{4.43e}$$

$$P(HD \leftarrow ISC) = \frac{1}{1 + \exp[-(-C_r - 1 + SB/f_C - 0)/\kappa]}, \tag{4.43f}$$

$$P(ID \leftarrow HSC) = \frac{1}{1 + \exp[-(-C_r + SB/f_C - (-1))/\kappa]}, \tag{4.43g}$$

$$P(ID \leftarrow ISC) = \frac{1}{1 + \exp[-(-C_r - 1 + SB/f_C - (-1))/\kappa]}. \tag{4.43h}$$

SB-RA

We refer to Eq. (4.15). The transition probabilities that we must consider are as follows:

$$P(HC \leftarrow D) = \frac{1}{1 + \exp[-(\langle\pi_D\rangle - (-C_r))/\kappa]}, \tag{4.44a}$$

$$P(IC \leftarrow D) = \frac{1}{1 + \exp[-(\langle\pi_D\rangle - (-C_r - 1))/\kappa]}, \tag{4.44b}$$

$$P(HSC \leftarrow D) = \frac{1}{1 + \exp[-(\langle\pi_D\rangle - 0)/\kappa]}, \tag{4.44c}$$

$$P(ISC \leftarrow D) = \frac{1}{1 + \exp[-(\langle\pi_D\rangle - (-1))/\kappa]}. \tag{4.44d}$$

$$P(HD \leftarrow C) = \frac{1}{1 + \exp[-(\langle \pi_C \rangle - 0)/\kappa]}, \tag{4.44e}$$

$$P(ID \leftarrow C) = \frac{1}{1 + \exp[-(\langle \pi_C \rangle - (-1))/\kappa]}, \tag{4.44f}$$

$$P(HSD \leftarrow C) = \frac{1}{1 + \exp[-(\langle \pi_C \rangle - 0)/\kappa]}, \tag{4.44g}$$

$$P(ISD \leftarrow C) = \frac{1}{1 + \exp[-(\langle \pi_C \rangle - (-1))/\kappa]}. \tag{4.44h}$$

For Model B, we should consider

$$P(HSC \leftarrow D) = \frac{1}{1 + \exp[-(\langle \pi_D \rangle - (-C_r + SB/f_C))/\kappa]}, \tag{4.45a}$$

$$P(ISC \leftarrow D) = \frac{1}{1 + \exp[-(\langle \pi_D \rangle - (-C_r - 1 + SB/f_C))/\kappa]}, \tag{4.45b}$$

$$P(HD \leftarrow C) = \frac{1}{1 + \exp[-(\langle \pi_C \rangle - 0)/\kappa]}, \tag{4.45c}$$

$$P(ID \leftarrow C) = \frac{1}{1 + \exp[-(\langle \pi_C \rangle - (-1))/\kappa]}. \tag{4.45d}$$

4.2.3.5 Global Time Evolution

Strategy updating takes place after each epidemic season as defined above (see Fig. 4.18). This inevitably brings increasing or decreasing x via increasing or decreasing f_C. Since there are four subsidy models and two strategy-updating rules, we can deduce eight different dynamical equations for predicting global FES as below.

Model A-1 + IB-RA

$$
\begin{aligned}
\frac{df_C}{dt} = & f_C(1-f_C)(1-\sigma')^2(e+(1-e)\exp[-R_0R(x,\infty)])\exp[-R_0R(x,\infty)] \\
& (P(HD \leftarrow HC) - P(HC \leftarrow HD)) \\
& + f_C(1-f_C)(1-\sigma')^2(e+(1-e)\exp[-R_0R(x,\infty)])(1-\exp[-R_0R(x,\infty)]) \\
& (P(ID \leftarrow HC) - P(HC \leftarrow ID)) \\
& + f_C(1-f_C)\sigma'(1-\sigma')(e+(1-e)\exp[-R_0R(x,\infty)])^2 \\
& (P(HSD \leftarrow HC) - P(HC \leftarrow HSD)) \\
& + f_C(1-f_C)\sigma'(1-\sigma')(1-e)(e+(1-e)\exp[-R_0R(x,\infty)]) \\
& (1-\exp[-R_0R(x,\infty)])(P(ISD \leftarrow HC) - P(HC \leftarrow ISD)) \\
& + f_C(1-f_C)(1-\sigma')^2(1-e)(1-\exp[-R_0R(x,\infty)])\exp[-R_0R(x,\infty)] \\
& (P(HD \leftarrow IC) - P(IC \leftarrow HD)) \\
& + f_C(1-f_C)(1-\sigma')^2(1-e)(1-\exp[-R_0R(x,\infty)])^2 \\
& (P(ID \leftarrow IC) - P(IC \leftarrow ID)) \\
& + f_C(1-f_C)\sigma'(1-\sigma')(1-e)(e+(1-e)\exp[-R_0R(x,\infty)]) \\
& (1-\exp[-R_0R(x,\infty)])(P(HSD \leftarrow IC) - P(IC \leftarrow HSD)) \\
& + f_C(1-f_C)\sigma'(1-\sigma')(1-e)^2(1-\exp[-R_0R(x,\infty)])^2 \\
& (P(ISD \leftarrow IC) - P(IC \leftarrow ISD)) \\
& + f_C(1-f_C)\sigma'(1-\sigma')(e+(1-e)\exp[-R_0R(x,\infty)])\exp[-R_0R(x,\infty)] \\
& (P(HD \leftarrow HSC) - P(HSC \leftarrow HD)) \\
& + f_C(1-f_C)\sigma'(1-\sigma')(e+(1-e)\exp[-R_0R(x,\infty)])(1-\exp[-R_0R(x,\infty)]) \\
& (P(ID \leftarrow HSC) - P(HSC \leftarrow ID)) \\
& + f_C(1-f_C)\sigma'^2(e+(1-e)\exp[-R_0R(x,\infty)])^2 \\
& (P(HSD \leftarrow HSC) - P(HCS \leftarrow HSD)) \\
& + f_C(1-f_C)\sigma'^2(1-e)(e+(1-e)\exp[-R_0R(x,\infty)])(1-\exp[-R_0R(x,\infty)]) \\
& (P(IDS \leftarrow HSC) - P(HCS \leftarrow ISD)) \\
& + f_C(1-f_C)\sigma'(1-\sigma')(1-e)(1-\exp[-R_0R(x,\infty)])\exp[-R_0R(x,\infty)] \\
& (P(HD \leftarrow ISC) - P(ISC \leftarrow HD)) \\
& + f_C(1-f_C)\sigma'(1-\sigma')(1-e)(1-\exp[-R_0R(x,\infty)])^2 \\
& (P(ID \leftarrow ISC) - P(ISC \leftarrow ID)) \\
& + f_C(1-f_C)\sigma'^2(1-e)(e+(1-e)\exp[-R_0R(x,\infty)])(1-\exp[-R_0R(x,\infty)]) \\
& (P(HDS \leftarrow ISC) - P(ICS \leftarrow HSD)) \\
& + f_C(1-f_C)\sigma'^2(1-e)^2(1-\exp[-R_0R(x,\infty)])^2 \\
& (P(ISD \leftarrow ISC) - P(ISC \leftarrow ISD)).
\end{aligned}
\tag{4.46}
$$

Model A-1 + SB-RA

$$\begin{aligned}\frac{df_C}{dt} = &-f_C(1-f_C)(1-\sigma')(e+(1-e)\exp[-R_0R(x,\infty)])P(HC \leftarrow D)\\&-f_C(1-f_C)(1-\sigma')(1-e)(1-\exp[-R_0R(x,\infty)])P(IC \leftarrow D)\\&-f_C(1-f_C)\sigma'(1-e)\exp[-R_0R(x,\infty)]P(HSC \leftarrow D)\\&-f_C(1-f_C)\sigma'(1-e)(1-\exp[-R_0R(x,\infty)])P(ISC \leftarrow D)\\&+f_C(1-f_C)(1-\sigma')\exp[-R_0R(x,\infty)]P(HD \leftarrow C)\\&+f_C(1-f_C)(1-\sigma')(1-\exp[-R_0R(x,\infty)])P(ID \leftarrow C)\\&+f_C(1-f_C)\sigma'(e+(1-e)\exp[-R_0R(x,\infty)])P(HSD \leftarrow C)\\&+f_C(1-f_C)\sigma'(1-e)(1-\exp[-R_0R(x,\infty)])P(ISD \leftarrow C).\end{aligned} \tag{4.47}$$

Model A-2 + IB-RA

$$\begin{aligned}\frac{df_C}{dt} = &f_C(1-f_C)(1-\sigma')(e+(1-e)\exp[-R_0R(x,\infty)])\exp[-R_0R(x,\infty)]\\&(P(HD \leftarrow HC)-P(HC \leftarrow HD))\\&+f_C(1-f_C)(1-\sigma')(e+(1-e)\exp[-R_0R(x,\infty)])(1-\exp[-R_0R(x,\infty)])\\&(P(ID \leftarrow HC)-P(HC \leftarrow ID))\\&+f_C(1-f_C)\sigma'(e+(1-e)\exp[-R_0R(x,\infty)])^2\\&(P(HSD \leftarrow HC)-P(HC \leftarrow HSD))\\&+f_C(1-f_C)\sigma'(e+(1-e)\exp[-R_0R(x,\infty)])(1-e)(1-\exp[-R_0R(x,\infty)])\\&(P(ISD \leftarrow HC)-P(HC \leftarrow ISD))\\&+f_C(1-f_C)(1-\sigma')(1-e)(1-\exp[-R_0R(x,\infty)])\exp[-R_0R(x,\infty)]\\&(P(HD \leftarrow IC)-P(IC \leftarrow HD))\\&+f_C(1-f_C)(1-\sigma')(1-e)(1-\exp[-R_0R(x,\infty)])(1-\exp[-R_0R(x,\infty)])\\&(P(ID \leftarrow IC)-P(IC \leftarrow ID))\\&+f_C(1-f_C)\sigma'(1-e)(1-\exp[-R_0R(x,\infty)])(e+(1-e)\exp[-R_0R(x,\infty)])\\&(P(ID \leftarrow IC)-P(IC \leftarrow ID))\\&+f_C(1-f_C)\sigma'(1-e)^2(1-\exp[-R_0R(x,\infty)])^2(P(ID \leftarrow IC)-P(IC \leftarrow ID)).\end{aligned} \tag{4.48}$$

Model A-2 + SB-RA

$$\begin{aligned}\frac{df_C}{dt} = &-f_C(1-f_C)(e+(1-e)\exp[-R_0R(x,\infty)])P(HC \leftarrow D)\\ &-f_C(1-f_C)(1-e)(1-\exp[-R_0R(x,\infty)])P(IC \leftarrow D)\\ &+f_C(1-f_C)(1-\sigma')\exp[-R_0R(x,\infty)]P(HD \leftarrow C)\\ &+f_C(1-f_C)(1-\sigma')(1-\exp[-R_0R(x,\infty)])P(ID \leftarrow C)\\ &+f_C(1-f_C)\sigma'(e+(1-e)\exp[-R_0R(x,\infty)])P(HSD \leftarrow C)\\ &+f_C(1-f_C)\sigma'(1-e)(1-\exp[-R_0R(x,\infty)])P(ISD \leftarrow C).\end{aligned} \tag{4.49}$$

Model A-3 + IB-RA

$$\begin{aligned}\frac{df_C}{dt} = &f_C(1-f_C)(1-\sigma')(e+(1-e)\exp[-R_0R(x,\infty)])\exp[-R_0R(x,\infty)]\\ &(P(HD \leftarrow HC) - P(HC \leftarrow HD))\\ &+f_C(1-f_C)(1-\sigma')(e+(1-e)\exp[-R_0R(x,\infty)])(1-\exp[-R_0R(x,\infty)])\\ &(P(ID \leftarrow HC) - P(HC \leftarrow ID))\\ &+f_C(1-f_C)(1-\sigma')(1-e)(1-\exp[-R_0R(x,\infty)])\exp[-R_0R(x,\infty)]\\ &(P(HD \leftarrow IC) - P(IC \leftarrow HD))\\ &+f_C(1-f_C)(1-\sigma')(1-e)(1-\exp[-R_0R(x,\infty)])(1-\exp[-R_0R(x,\infty)])\\ &(P(ID \leftarrow IC) - P(IC \leftarrow ID))\\ &+f_C(1-f_C)\sigma'(e+(1-e)\exp[-R_0R(x,\infty)])\exp[-R_0R(x,\infty)]\\ &(P(HD \leftarrow HSC) - P(HSC \leftarrow HD))\\ &+f_C(1-f_C)\sigma'(e+(1-e)\exp[-R_0R(x,\infty)])(1-\exp[-R_0R(x,\infty)])\\ &(P(ID \leftarrow HSC) - P(HSC \leftarrow ID))\\ &+f_C(1-f_C)\sigma'(1-e)(1-\exp[-R_0R(x,\infty)])\exp[-R_0R(x,\infty)]\\ &(P(HD \leftarrow ISC) - P(ISC \leftarrow HD))\\ &+f_C(1-f_C)\sigma'(1-e)(1-\exp[-R_0R(x,\infty)])(1-\exp[-R_0R(x,\infty)])\\ &(P(ID \leftarrow ISC) - P(ISC \leftarrow ID)).\end{aligned} \tag{4.50}$$

Model A-3 + SB-RA

$$
\begin{aligned}
\frac{df_C}{dt} = & -f_C(1-f_C)(1-\sigma')(e+(1-e)\exp[-R_0R(x,\infty)])P(HC \leftarrow D) \\
& -f_C(1-f_C)(1-\sigma')(1-e)(1-\exp[-R_0R(x,\infty)])P(IC \leftarrow D) \\
& -f_C(1-f_C)\sigma'(e+(1-e)\exp[-R_0R(x,\infty)])P(HSC \leftarrow D) \\
& -f_C(1-f_C)\sigma'(1-e)(1-\exp[-R_0R(x,\infty)])P(ISC \leftarrow D) \\
& +f_C(1-f_C)\exp[-R_0R(x,\infty)]P(HD \leftarrow C) \\
& +f_C(1-f_C)(1-\exp[-R_0R(x,\infty)])P(ID \leftarrow C).
\end{aligned} \tag{4.51}
$$

Model B + IB-RA

$$
\begin{aligned}
\frac{df_C}{dt} = & f_C(1-f_C)(e+(1-e)\exp[-R_0R(x,\infty)])\exp[-R_0R(x,\infty)] \\
& (P(HD \leftarrow HSC) - P(HSC \leftarrow HD)) \\
& +f_C(1-f_C)(e+(1-e)\exp[-R_0R(x,\infty)])(1-\exp[-R_0R(x,\infty)]) \\
& (P(ID \leftarrow HSC) - P(HSC \leftarrow ID)) \\
& +f_C(1-f_C)(1-e)(1-\exp[-R_0R(x,\infty)])\exp[-R_0R(x,\infty)] \\
& (P(DH \leftarrow ISC) - P(ISC \leftarrow HD)) \\
& +f_C(1-f_C)(1-e)(1-\exp[-R_0R(x,\infty)])^2 \\
& (P(ID \leftarrow ISC) - P(ISC \leftarrow ID)).
\end{aligned} \tag{4.52}
$$

Model B + SB-RA

$$
\begin{aligned}
\frac{df_C}{dt} = & -f_C(1-f_C)(e+(1-e)\exp[-R_0R(x,\infty)])P(HSC \leftarrow D) \\
& -f_C(1-f_C)(1-e)(1-\exp[-R_0R(x,\infty)])P(ISC \leftarrow D) \\
& +f_C(1-f_C)\exp[-R_0R(x,\infty)]P(HD \leftarrow C) \\
& +f_C(1-f_C)(1-\exp[-R_0R(x,\infty)])P(ID \leftarrow VC).
\end{aligned} \tag{4.53}
$$

All dynamical equations above can be solved numerically. We introduce a so-called explicit scheme for the time-varying terms to obtain a numerical solution for vaccination coverage at equilibrium.

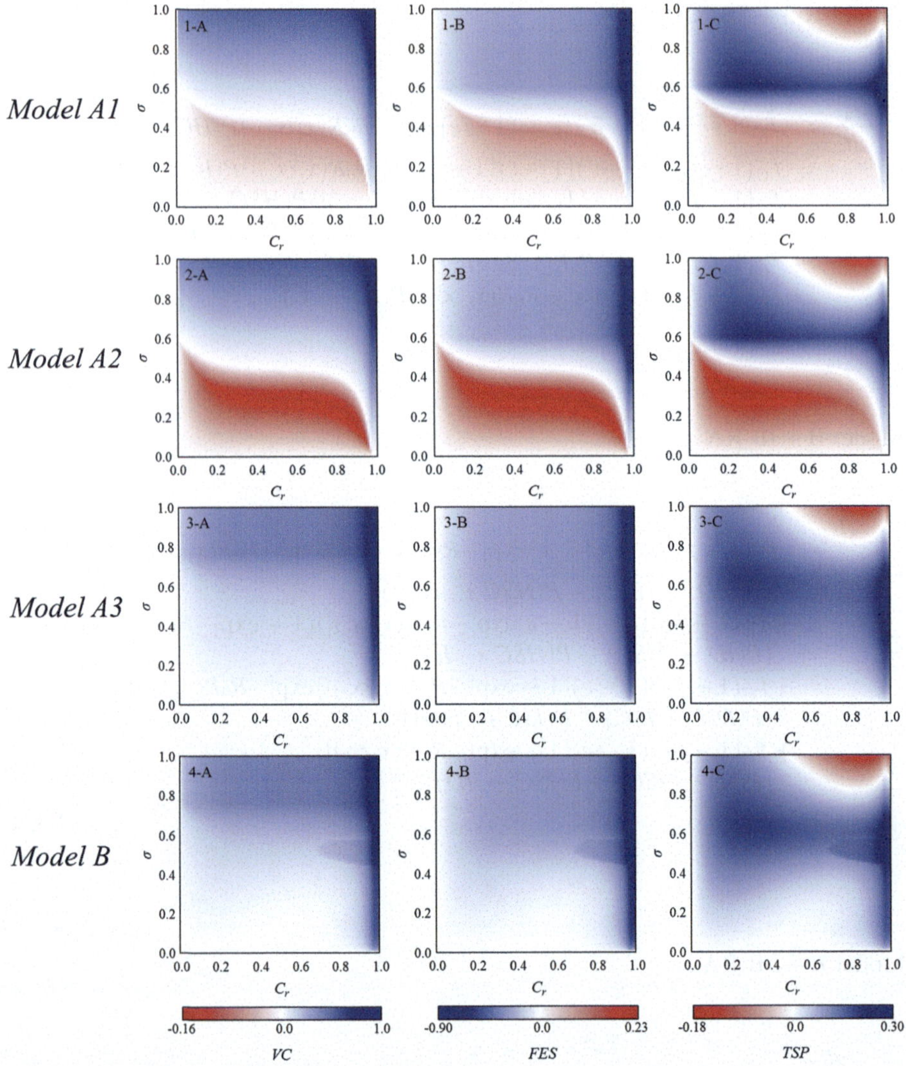

Fig. 4.29 Analytical result: Color indicates difference between subsidy ($\sigma > 0$) and non-subsidy ($\sigma = 0$) cases. Panel (*-A) Vaccination coverage (VC), (*-B) final epidemic size (FES), (*-C) total social payoff (TSP). Panel (1-*) Model A1, (2-*) Model A2, (3-*) Model A3, (4-*) Model B. $e = 1.0$ and IB-RA are presumed

4.2.3.6 Discussion

Figures 4.29, 4.30, 4.31, and 4.32 present our analytical results. Figures 4.29 and 4.30 provide the results under the four types of subsidy policies when respectively presuming $e = 1.0$ (perfect vaccination) and $e = 0.4$ (low reliable vaccination) using IB-RA as a strategy-updating rule. Figures 4.31 and 4.32 present the same presuming SB-RA.

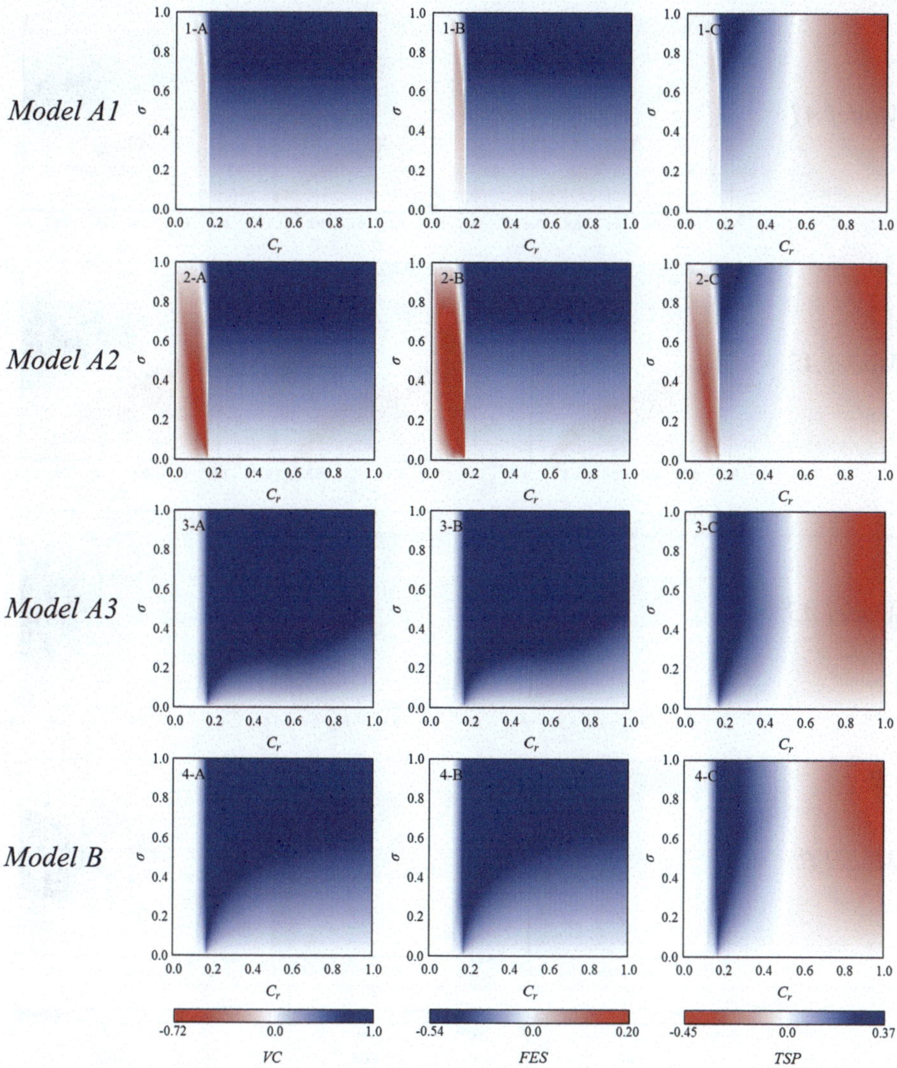

Fig. 4.30 Analytical result: Color indicates difference between subsidy ($\sigma > 0$) and non-subsidy ($\sigma = 0$) cases. Panel (*-A) Vaccination coverage (VC), (*-B) final epidemic size (FES), (*-C) total social payoff (TSP). Panel (1-*) Model A1, (2-*) Model A2, (3-*) Model A3, (4-*) Model B. $e = 0.4$ and IB-RA are presumed

Because of the mean-field approximation, panels (VC, FES and TSP) in the top row (assuming Model A1) in Fig. 4.29 should be compared with Fig. 4.24 presuming RRG. Qualitatively, both show the same tendency. Differences in detail level come from finite resolution and insufficient population, as well as insufficient average degree in the simulation, which seems to some extent inevitable. Thus, we conclude that our theoretical model well-reproduces the MAS-simulation result.

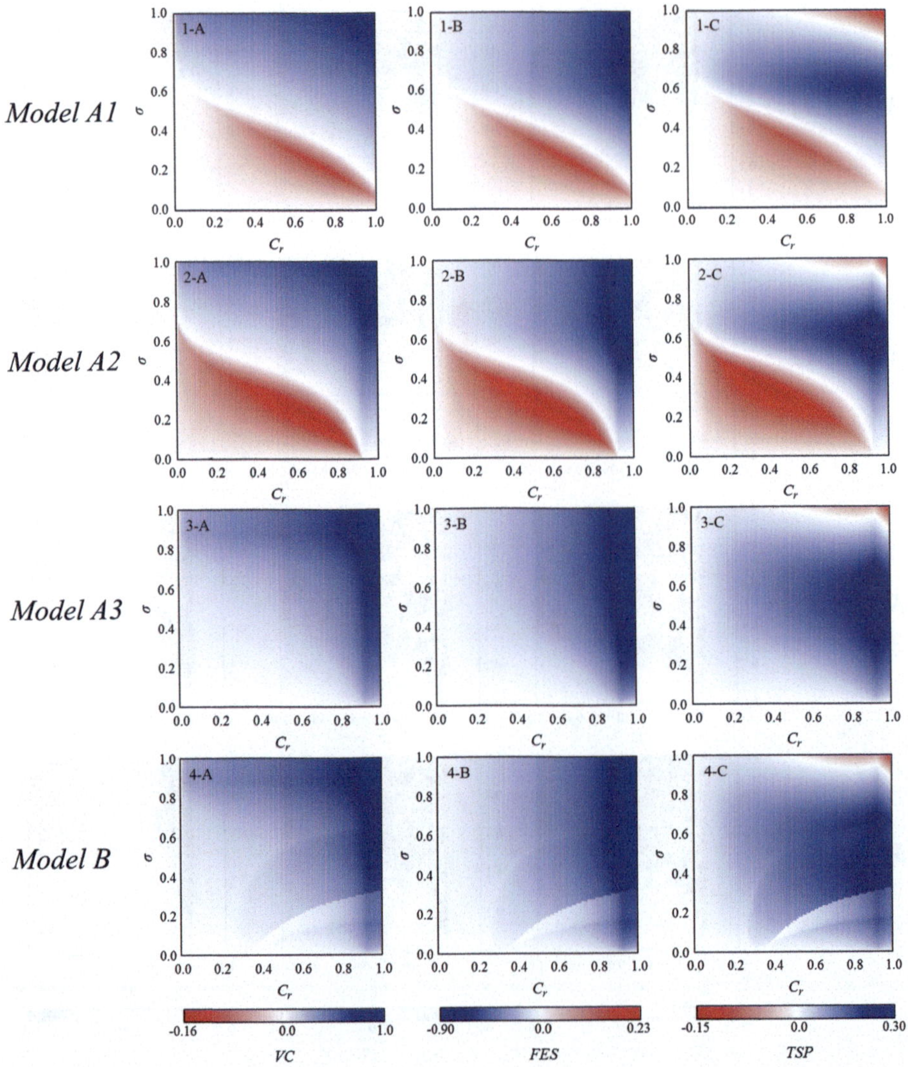

Fig. 4.31 Analytical result: Color indicates difference between subsidy ($\sigma > 0$) and non-subsidy ($\sigma = 0$) cases. Panel (*-A) Vaccination coverage (VC), (*-B) final epidemic size (FES), (*-C) total social payoff (TSP). Panel (1-*) Model A1, (2-*) Model A2, (3-*) Model A3, (4-*) Model B. $e = 1.0$ and SB-RA are presumed

Let us compare the subsidy systems assuming perfect vaccination ($e = 1.0$) and IB-RA in Fig. 4.29. Focusing on TSP under Models A-1 and A-2, the negative region exists in a relatively smaller subsidy size of σ(hereafter, the first negative region) and a larger subsidy size of σ, dovetailed with a higher vaccination cost (hereafter, second negative region). In the first negative region, a subsidy going to

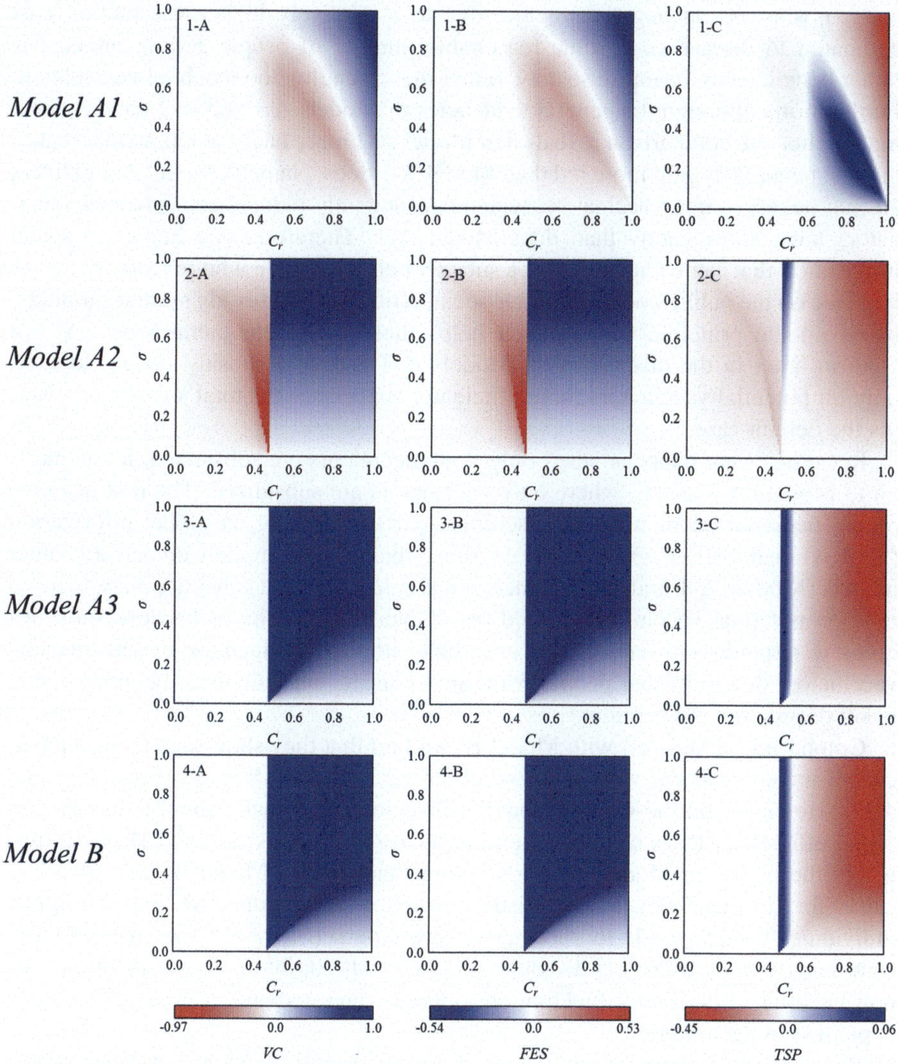

Fig. 4.32 Analytical result: Color indicates difference between subsidy ($\sigma > 0$) and non-subsidy ($\sigma = 0$) cases. Panel (*-A) Vaccination coverage (VC), (*-B) final epidemic size (FES), (*-C) total social payoff (TSP). Panel (1-*) Model A1, (2-*) Model A2, (3-*) Model A3, (4-*) Model B. $e = 0.4$ and SB-RA are presumed

non-vaccinators (i.e., defectors) counterproductively works to suppress the social cost because being a defector becomes cost-advantageous (since they may avoid infection by being given free-tickets), reducing the fraction of cooperators (i.e., vaccinators) as compared with the default case. As a consequence, the number of self-financed vaccinators comes down, worsening the social efficiency. On the other hand, in the second negative region, due to a relatively higher vaccination cost compared to disease cost, some reasonable number of people getting infected is rather beneficial to the entire society, rather than spending too much on vaccination. Furthermore, although the general tendencies of Models A-1 and A-2 are the same, more deliberate comparison reveals that Model A-2 more badly works so that makes the first negative region more red than Model A-1 does. Namely, Model A-2 delivers all free-tickets to non-vaccinators, which consequently reduces self-financed vaccinators more significantly than does Model A-1. Therefore, one important social implication that can be noted is that a subsidy helping people who potentially aim to free-ride on the public good devastates social efficiently. A subsidy system should be based on the principle that "heaven helps those who help themselves". As we confirm latter in the discussion on Models A-3 and B, a subsidy system focused only on potential vaccinators more efficiently suppresses the total social cost vis-a-vis the default case.

Recalling what happens when only non-vaccinators are subsidized, let us move on to Models A-3 and B, where only vaccinators are subsidized. The first negative region does not occur under any settings, whether relying on global information (SB-RA) or not (IB-RA). As a whole, those two subsidy models outperform other models (Models A-1 and A-2). Hence, we would say that a subsidy policy focused only on potential vaccinators should be adopted. Again, this is because subsidies going to potential non-vaccinators eventually impede the increase of self-financed vaccinators due to misled people who aim either to free-ride or to be given a free ticket despite not cooperating.

Comparing Model A-3 with Model B, we find that they show analogous tendencies. This is consistent with what we observed in the MAS result (Figs. 4.21 and 4.22). However, on the whole. Model A-3 seems better than Model B in terms of social efficiency. Pairs of the broken-line boxes in Models A-3 and B prove this, where the region in Model A-3 looks bluer than that by Model B. One plausible cause for this tendency is that distributing free tickets only to limited eligible individuals drives people to vaccinate (increasing self-financed vaccinators) more significantly than offering a discount coupon to all eligible individuals, due to the non-linearity of the Fermi function considered when updating a strategy. This can explained in detail below.

Recall the two arguments (i.e., payoffs) in the Fermi function. Let us suppose a non-vaccinator (defector) copies V from a vaccinator (cooperator) who is given either a free-ticket in the case of Model A-3 or a discount coupon in the case if Model B, and compare those two models. Note that the state-transition probability of Model A-3 is larger than that of Model B, which is attributed to the reduction in price by a free-ticket or discount-coupon. Since the same σ is presumed for this comparison, the fraction (number) of vaccinators given free-tickets in Model A-3 is less than that given discount-coupons (that is consistent with the entire number of vaccinators) in Model B, which is considered in the final dynamical equations (Eq. (4.50) or Eq. (4.51) for Model A-3 and Eq. (4.52) or Eq. (4.53) for Model B). But, again, because of non-linearity of the Fermi function (state-transition probability), the attractive force causing NV to become V in Model A-3 is greater than that in Model B.

Comparing Fig. 4.29 with Fig. 4.30, or Fig. 4.31 with Fig. 4.32, with decrease of vaccine reliability (decrease of e) the first negative region becomes smaller whereas the second negative region is larger. Interestingly, in the case assuming $e = 0.4$ and IB-RA, both regions in which introducing a subsidy is not justified are insensitive to σ. In fact, if C_r becomes more than 0.6, the subsidy obviously deteriorates the social efficiency regardless of the size of subsidy. For C_r below 0.2, this deterioration effect becomes more slight.

Figures 4.31 and 4.32, presuming SB-RA, present a quite different picture than Figs. 4.29 and 4.30, resulting from the difference in the strategy-update rules. Let us compare our respective TSPs with $e = 0.4$. The black and gray boxes indicate the first and second negative regions. As discussed above, the first negative region results from the situation whereby subsidizing non-vaccinators hampers the increase of self-financed vaccinators, while the second region is brought about by the fact that spending too much on vaccination becomes less beneficial on the whole than allowing a reasonable level of infectious individuals. Although the first negative regions (black boxes) are at comparable levels, the second negative region (gray box) of Fig. 4.32 is less than that of Fig. 4.30, only appearing at larger $\sigma$and larger C_r. Thus, sharing global information during strategy-updating events (SB-RA) helps to justify a subsidy system. Observing carefully, we can note that the range of vaccination costs justifying subsidies (colored blue) is insensitive to σ; however, although it clearly appears around $0.2 \leq C_r \leq 0.5$ in Fig. 4.30, it almost disappears in Fig. 4.32. If we make the same comparison for Model A-2 in Figs. 4.30 and 4.32 (see black and gray boxes in respective right panels), we find the same behavior. There is none of any range of vaccination cost justifying subsidy insensitive to σ in Model A-2 of Figs. 4.31 and 4.32. Moreover, remarkably, in the right-hand panel of Model A-2 in Fig. 4.32, there is almost no parameter region in which a subsidy is justified. Subsidizing only non-vaccinators (Model A-2) in the case presuming a strategy-updating rule relying on global information (SB-RA) and a unreliable vaccination ($e = 0.4$) is not justified at all.

4.2.4 Summary and Social Implications

In order to help society establish an effective subsidy policy to combat the spread of infectious disease and mitigate the risk of pandemics, this study proposed a comprehensive "vaccination game", wherein a subsidy system is considered in the context of both the dynamics of individual decision-making based on evolutionary game theory and the spread of disease using the SIR/V model through a social network with consideration of a subsidy system.

For our analysis, we performed not only multi-agent simulation (MAS) considering how the underlying topology of the social network affected equilibrium, but also a theoretical approach presuming a mean-field approximation. In particular, our analytic model deals with imperfectly working vaccination parameterized by "effectiveness", which does not always bring a perfect immunity by taking a vaccine.

We presume four types of subsidy systems depending on whether a free-ticket or a discount-coupon is given, as well as individual attributes, such as being a potential vaccinator or a defector trying to free-ride on herd immunity.

We mainly observed our results in terms of vaccination coverage (VC), final epidemic size (FES), and total social payoff (TSP) (or, looking negatively, total social cost), using these to indicate social efficiency.

First of all, we confirmed that our analytical approach is capable of reproducing the result obtained by the MAS approach.

Our result suggests that spending too little on subsidy or too much for a relatively higher vaccination cost results in an ironic situation where introducing a subsidy incurs a higher social cost than the default case. Little spending on the subsidy results in making self-financed vaccinators decrease as a fraction of society (hereafter; let us call this the "first regime"). Overspending on a subsidy when the vaccination cost is high brings the situation that rather some people being infected becomes rather socially efficient than too much vaccinators due to the relation of vaccination cost to infection cost; this devastates the social efficiency compared to the default case realizing (which we hereafter call the "second regime").

The MAS result shows that the underlying social network significantly influences equilibrium. In particular, a scale-free network rather than a lattice expands the parameter region in which a subsidy system deteriorates the social efficiency.

If a vaccine's reliability degrades (presuming low effectiveness), the parameter region in which a subsidy is counterproductive due to the second regime grows and becomes less sensitive to the subsidy size.

A subsidy that applies only to potential cooperators is quite important for the optimal social design of a subsidy system. Although a subsidy applying to people who have no intention of vaccinating unless given either a free-ticket or a discount-coupon might be thought efficient, or at least socially favored or accepted in the context of a high-welfare society, such a scheme could reduce the number of inherently cooperative vaccinators (self-financed vaccinators), owing to disregard for the principle of "heaven helping those who help themselves".

Although the difference between free-ticket policy and discount-coupon policy was observed to be small (so long as tickets were only given to potential vaccinators), this theoretical approach shows that the free-ticket slightly outperforms the discount-coupon policy. This is because a larger payoff difference is brought by a free-ticket instead of a discount-coupon, triggering an increase in self-financed vaccinators.

References

Amaral, M. A.; Wardil, L., Perc, M., da Silva, J. K. L.; Stochastic win-stay-lose-shift strategy with dynamic aspirations in evolutionary social dilemmas, *Physical Review E* **94**, 032317, 2016.

Anderson, R.M., May, R. M.; Infectious diseases of humans, Oxford University Press, 1991.

Asch, D. A., Baron, J., Hershey, J. C, Kunreuther, H., Meszaros, J., Ritov, I., Spranca, M.; Omission bias and pertussis vaccination, *Medical Decision Making* **14**, 118–123, 1994.

Axelrod, R.; An evolutionary approach to norms, *American Political Science Review* **80 (4)**, 1095–1111, 1986.

Bai, F.; Uniqueness of Nash equilibrium in vaccination games, *Journal of Biological Dynamics* **10 (1)**, 395–415, 2016.

Barabási, A. L., Albert, R.; Emergence of scaling in random networks. *Science* **286**, 509–512, 1999.

Basu, S., Chapman, G. B., Galvani, A. P.; Integrating epidemiology, psychology, and economics to achieve HPV vaccination targets, *Proceedings of the National Academy of Science of the United States of America* **105**, 19018–19023, 2008.

Bauch, C. T.; Imitation dynamics predict vaccinating behavior. *Proceedings of the Royal Society B* **272**, 1669–1675, 2005.

Bauch, C. T., Earn, D. J. D.; Vaccination and the theory of games, *Proceedings of the National Academy of Science of the United States of America* **101**, 13391–13394, 2004.

Bauch, C. T., Galvani, A. P., Earn, D. J. D.; Group interest versus self interest in smallpox vaccination policy, *Proceedings of the National Academy of Science of the United States of America* **100**, 10564–10567, 2003.

Bollobás, B., Random graphs, Academic Press, London, 1985.

Brian Arthur, W.; Inductive Reasoning and Bounded Rationality, *American Economic Review* **84**, 406–411, 1994.

Cardillo, A., Reyes-Suárez, C., Naranjo, F., & Gómez-Gardeñes, J.; Evolutionary vaccination dilemma in complex networks, *Physical Review E* **88**, 032803, 2013.

Challet, D., Marsili, M., Zhang, Y.-C.; Minority Games: Interacting Individuals in Financial Markets, Oxford University Press, 2005.

Chapman, G. B., Coups, E. J.; Predictors of influenza vaccine acceptance among healthy adults, *Preventive Medicine* **29 (4)**, 249–262, 1999.

Chapman, G. B., Coups, E. J.; Emotions and preventive health behavior: worry, regret, and influenza vaccination, *Health Psychology* **25**, 82–90, 2006.

Chen, X., Wang, L.; Promotion of cooperation induced by appropriate payoff aspirations in a small-world networked game, *Physical Review E* **77**, 017103, 2008.

Cullen, J., West, P.; The economics of health. An introduction. In Martin Robertson, Oxford: Martin Robertson, 1979.

Ding, H., Xu, J.-H., Wang, Z., Ren, Y.-Z., Cui, G.-H.; Subsidy strategy based on history information can stimulate voluntary vaccination behaviors on seasonal diseases, *Physica A* **503**, 390–399, 2018.

Fine, P., Clarkson, J.; Individual versus public priorities in the determination of optimal vaccination policies, *American Journal of Epidemiology* **124**, 1012–1020, 1986.

Fu, F., Rosenbloom, D. I., Wang, L., Nowak, N. A.; Imitation dynamics of vaccination behavior on social networks, *Proceedings of the Royal Society B* **278**, 42–49, 2011.

Fukuda, E., Tanimoto, J.; Effects of stubborn decision-makers on vaccination and disease propagation in social networks, *International Journal of Automation and Logistics* **2**, 78–92, 2016.

Fukuda, E., Kokubo, S., Tanimoto, J., Wang, Z., Hagishima, A., Ikegaya, N.; Risk assessment for infectious disease and its impact on voluntary vaccination behavior in social networks, *Chaos, Solitons & Fractals* **68**, 1–9, 2014.

Fukuda, E., Tanimoto, J., Akimoto, M; Influence of breaking the symmetry between disease transmission and information propagation networks on stepwise decisions concerning vaccination, *Chaos, Solitons & Fractals* **80**, 47–55, 2015.

Gavious, A., Yamin, D.; Incentives 'effect in influenza vaccination policy, *Management Science* **59 (12)**, 2667–2682, 2013.

Geoffard, P., Philipson, T.; Disease eradication: private versus public vaccination, *American Economic Review* **87**, 222–230, 1997.

Gillespie, D. T. J.; Exact stochastic simulation of coupled chemical reactions, *Journal of Physical Chemistry* **81**, 2340–2361, 1977.

Hethcote, H. W., van den Driessche, P; An SIS epidemic model with variable population size and a delay, Journal of Mathematical Biology **34**, 177–194, 1995.

Imhof, L. A., Fudenberg, D., Nowak, M. A.; Tit-for-Tat or win-stay, lose-shift?, *Journal of Theoretical Biology* **247**, 574–580, 2007.

Iwamura, Y, Tanimoto, J., Fukuda, E.; Effect of intermediate defense measures in voluntary vaccination games, *Journal of Statistical Mechanics: Theory and Experiment*, 093501, 2016.

Jansen, V. A., Stollenwerk, N., Jensen, H. J., Ramsay, M. E., Edmunds, W. J., Rhodes, C. J.; Measles outbreaks in a population with declining vaccine uptake. *Science* **301**, 804, 2003.

Keeling, M. J., Eames, K. T. D.; Networks and epidemic models, *Journal of the Royal Society Interface* **2**, 295–307, 2005.

Kermack, W. O., McKendrick, A. G.; A contribution to the mathematical theory of epidemics, *Proceedings of Royal Society of London, Series A*, 700–721, 1927.

Kuga, K., Tanimoto, J.; Which is more effective for suppressing an infectious disease: imperfect vaccination or intermediate defense measures?, *Journal of Statistical Mechanics: Theory and Experiment*, 023407, 2018.

Li, Q., Li, M.-C., Lv, L., Guo, C., Lu, K.; A new prediction model of infectious disease with vaccination strategies based on evolutionary game theory, *Chaos, Solitons & Fractals* **104**, 51–60, 2017.

Macy, M. W., Flache, A.,; Learning dynamics in social dilemmas, *Proceedings of the National Academy of Science of the United States of America* **99 (3)**, 7229–7236, 2002.

Masuda, N.; Temporal network epidemiology, Springer, 2017.

Matsuzawa, R., Tanimoto, J., Fukuda, E.; Spatial prisoner's dilemma games with zealous cooperators, *Physical Review E* **94**, 022114, 2016.

Olson, M.; The Logic of Collective Action, Cambridge University Press, 1965.

Pastor-Satorras, R., Vespignani, A.; Epidemic spreading in scale-free networks, *Physical Review Letters* **86**, 3200, 2001.

Sahimi, M.; Applications of Percolation Theory, Taylor & Francis, 1994.

Tang, G.-M., Cai, C.-R., Wu, Z.-X.; Evolutionary vaccination dynamics with internal support mechanisms, *Physica A* **473**, 135–143, 2017.

Tanimoto, J.; Mathematical Analysis of Environmental System, Springer, 2014.

Vardavas, R., Breban, R., Blower, S.; Can influenza epidemics be prevented by voluntary vaccination?, *PLoS Computation Biolpgy* **3 (5)**, e85, 2007.

Watts, D.J., Strogatz, S.H.; Collective dynamics of 'small-world' networks, *Nature* **393**, 440–442, 1998.

Wu B., Fu F. & Wang L.; Imperfect vaccine aggravates the long-standing dilemma of voluntary vaccination. *PLoS One* **6**, e20577, 2011.

Yamagishi, T.; The provision of a sanctioning system as a public good, *Journal of Personality and Social Psychology* **51**, 110–116, 1986.

Zhang, H.-F., Wu, Z.-X., Xu, X.-K., Small, M., Wang, L., Wang, B.-H.; Impact of subsidy policies on vaccination decisions in contact networks, *Physical Review E* **88**, 012813, 2013.

Zhang, H.-F., Wu, Z.-X., Tang, M., Loi, Y.-C.; Effects of behavioral response and vaccination policy on epidemic spreading – an approach based on evolutionary-game dynamics, *Scientific Reports*, 4:5666, 2014.

Zhang, H.-F., Shu, P.-P., Wang, Z., Tang, M.; Preferential imitation can invalidate targeted subsidy policies on seasonal-influenza diseases, *Applied Mathematics and Computation* **294**, 332–342, 2017.

Chapter 5
Past and Future: Evolutionary Game Theory

Throughout this book, we have described evolutionary game theory (EGT), which is one of the most powerful tools for modelling, analyzing, and predicting the complex dynamics of human—environmental–social systems. Subsequently, we discussed the successful application of EGT to actual human social systems. In particular, we have focused on studying traffic flow and epidemiology based on multi-agent-simulation (MAS) models. Physical dynamics, such as compressive-fluid-like traffic flow and the spread of an infectious disease dovetail with the framework of an evolutionary game representing the decision-making process of agents. This approach has resulted in some new findings.

To conclude this intellectual voyage, it would be valuable to ask ourselves whether EGT is a versatile "ultimate weapon" for investigating complex human systems.

The answer is: of course not. EGT offers nothing more than one narrow possibility. Real human systems, on the other hand, are so complex that they remain neither ascertainable nor predictable.

Let us present an interesting proof that questions EGT's universality.

As we have repeatedly noted in this book, Professor Nowak's milestone work[1] — which profoundly alleged that any mechanisms from which mutual cooperation emerges can be described in terms of adding "social viscosity" to an original system in which myriad "anonymous" agents compete—seems significantly persuasive.

From a biological standpoint, the most important bifurcation point distinguishing humans from other animal species and resulting in our greater prosperity is the emergence of language. Language is supreme vis-à-vis other primitive communication protocols in that it identifies others around oneself, and also helps to convey what one is thinking, wanting, and aiming for to one's neighbors. This first

[1]Nowak (2006).

J. Tanimoto, *Evolutionary Games with Sociophysics*, Evolutionary Economics and Social Complexity Science 17, https://doi.org/10.1007/978-981-13-2769-8_5

revolutionary invention has made humans the lords of the Earth. After the introduction of language, humans started to create tribal societies, resulting in an extreme decrease in anonymity compared with the primitive state of fragmented individuals; in other words, language added a great deal of social viscosity.

The second great revolution in human history was the introduction of agriculture. Some historians insist that the history of agriculture started 23,000-years ago around the Sea of Galilee. Growing out from a tribal social system, a number of prodigious kingdoms appeared, including Ancient Egypt, Mesopotamia, the Indus culture, and the Yellow River culture. Large-scale accumulation of food resources enabled humans to establish ancient civilizations and their many byproducts, including writing systems. This system allowed our ancestors to realize an elaborate social system relying on, for example, a legal system and religion. As compared with the tribal society in the previous era, more solid social viscosity was introduced. Our earlier discussion supports the idea that humans became more affluent by introducing a more fixed social viscosity.

The third epoch-making event in human history was the Industrial Revolution, which started in Britain in the late-eighteenth century and spread to other Western countries. However, we should note that the inception of the Industrial Revolution could be dated to the Renaissance, taking place in fourteenth century Italy.

Since the Industrial Revolution, world population has increased geometrically. So-called "modernism" originated in the Western world as a result of the Industrial Revolution and brought a blissful peak to human material and physical welfare. Although we have experienced many historical tragedies—colonization, large-scale warfare and genocide—material abundance has become a reality.

But now the tide has turned.

A new development—the information-technology (IT) revolution—has brought another wave of change in the post-modern era. It is worthwhile to point out that the relationship between social viscosity and social productivity is what it used to be, namely, when adding social viscosity raised productivity (which seems explicable from an EGT standpoint). In fact, the IT Revolution makes much higher efficiency possible in industrial fields, but it simultaneously enables an anonymous society, which makes us feel more comfortable in the meaning that urban life might be more convenient than traditional (perhaps; rural) life-style and free from many conventions binding people to solid communities. It seems bizarre that such an inverse relation should take place. If this turns out to be a real inverse relationship between social viscosity and mutual cooperation, evolutionary game theory—a unique means of quantifying human–environmental–social systems—must provide an explanation as to why (Fig. 5.1).

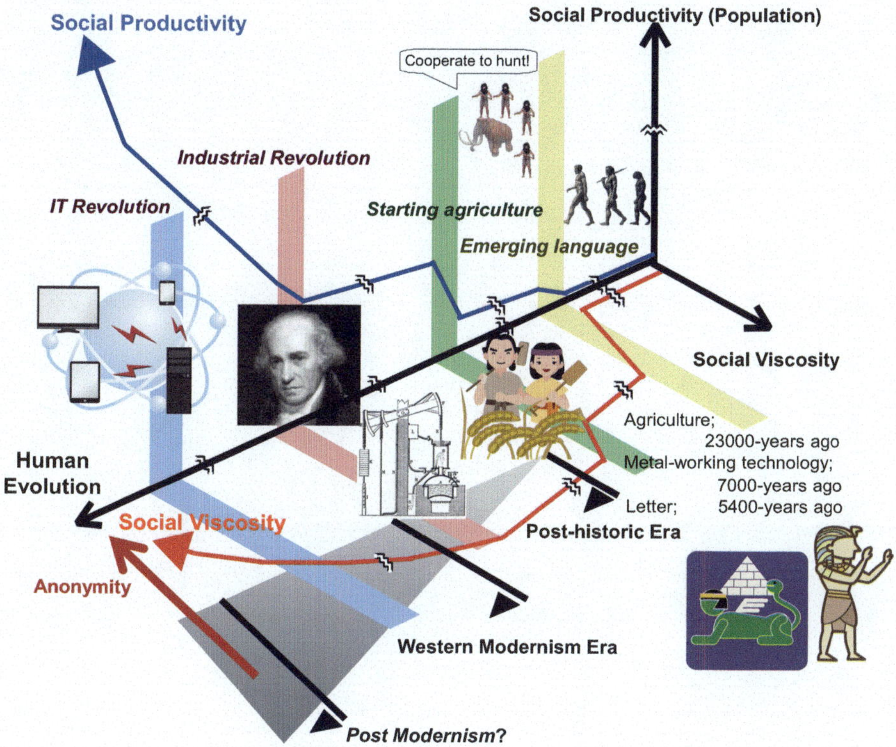

Fig. 5.1 Can we re-build a "Community" that supports the cooperative propensity that humanity may intrinsically have?

Reference

Nowak, M. A.; Five rules for the evolution of cooperation, *Science* 314, 1560–1563, 2006.

Index

J. Tanimoto, *Evolutionary Games with Sociophysics*, Evolutionary Economics and Social Complexity Science 17, https://doi.org/10.1007/978-981-13-2769-8

The manufacturer's authorised representative in the EU is Springer Nature Customer Service Centre GmbH, Europaplatz 3, 69115 Heidelberg, Germany. If you have any concerns regarding our products, please contact ProductSafety@springernature.com

Printed and bound by CPI Group (UK) Ltd, Croydon, CR0 4YY
25/09/2026
02222892-0006